The Disappearance of Insight

The Disappearance of Insight

When Technology Undermines How and Why

Frank C. Keil

The MIT Press
Cambridge, Massachusetts
London, England

The MIT Press
Massachusetts Institute of Technology
77 Massachusetts Avenue
Cambridge, MA 02139
mitpress.mit.edu

The MIT Press would like to thank the anonymous peer reviewers who provided comments on drafts of this book. The generous work of academic experts is essential for establishing the authority and quality of our publications. We acknowledge with gratitude the contributions of these otherwise uncredited readers.

This book was set in Stone Serif and Stone Sans by Westchester Publishing Services. Printed and bound in the United States of America.

Library of Congress Cataloging-in-Publication Data is available.

ISBN: 978-0-262-05414-0

10 9 8 7 6 5 4 3 2 1

EU Authorised Representative: Easy Access System Europe, Mustamäe tee 50, 10621 Tallinn, Estonia | Email: gpsr.requests@easproject.com

For Kristi

Contents

Preface

We naturally seek explanations when we encounter things we don't understand. Hummingbirds are the only bird that can fly backward. Tidal ranges are larger in more northern latitudes. Heat pumps cool and heat while air conditioners only cool. The mere mention of these cases triggers a desire to learn more about them. Even preverbal children seek explanations for phenomena by exploring and manipulating puzzling things until they understand them. The drive to know why and how, an active form of wondering, emerges early in childhood and is a source of delight. That drive supports our developing abilities to make sense of our surroundings and to build reliable communities of knowledge.

Young children have strong drives to know why and how, but those drives are sustained only when the world around them cooperates. Eventually, much of this worldly support may come from conversations with others, but early on, direct experience with physical things is critical. The pleasure of achieving insights often arises when exploration or tinkering reveals hidden components and causal patterns. But, with many artifacts and devices, these rewarding experiences are disappearing. Complete mechanistic explanations of modern devices are known by experts, but those explanations are no longer recoverable from an inspection of their parts. Entities are unexplainable in this sense because the relevant causal details have become inaccessible to nonexperts. This new opacity especially confronts young children who, even as infants, have a deep fascination with what is going on beneath the surface while being unable to learn through verbal explanations.

Wondering needs to be supported and nourished early on lest it rapidly wither in later years. Even when the drive to ask why and how is kept burning bright, it needs fertile fields in which to grow. Visual inspection and tinkering are no longer feasible ways to figure out the workings of common objects such as telephones, toasters, and kitchen scales. Those changes may influence the growth of children's cognitive abilities, as well as adults' understanding, in ways that extend far beyond artifacts.

Plants and animals do still provide some opportunities to explore and manipulate. We can see how Velcro was inspired by close inspection and manipulation of burdock burrs and how they attach to animals and clothing. An attentive and patient observer can, with a little luck, see how a spider spins a web more than twenty times its body size. Such revealing biological examples, however, are less common experiences for most children today than they were in the past.

A stronger mechanical sense of devices in children and lay adults may have traditionally helped build deeper understandings of nature. But massive, yet often unnoticed, changes in the world of artifacts have occurred—changes that are eliminating those paths to insight. Devices, tools, and instruments are central to how we transmit and elaborate on knowledge as a culture. The story with artifacts then suggests broader implications relating to divisions of expertise in successful cultures and how we understand all that is around us.

In a different way, causal explanations and the pleasures of insight may be disappearing for people of all ages and all levels of expertise. While experts may be deriving ever more accurate predictions, they find themselves striking Faustian bargains by giving up explanatory understanding. But perhaps those bargains are benign. Even if we can't understand massive weather forecast computations, maybe all we need are reliable predictions. If the latest weather model reliably tells us when a cold snap is coming, do we need to also know how that model works? In contrast to how science has typically proceeded for centuries, greater predictive ability no longer implies greater explanatory power. Building predictive machines often entails losing a sense of how those machines generate their results. Such trade-offs may be a necessary part of many advanced technologies and sciences; in some areas, younger generations may value the prediction side of the trade-off more than earlier generations that favored explanations. New technologies shouldn't be condemned just because parts of them are conceptually impenetrable. Instead of raging against these new machines, we need to examine the consequences of not knowing how they work. In some cases, our ignorance may be neatly bounded and managed—but in other situations, we may struggle to know what we don't know and should be aware of that struggle and the risks it poses.

While these two senses of diminishing insights are starkly different, young children, in their successes and failures, suggest ways to meet the challenges posed by both versions. As they enter formal schooling, children use a small number of powerful "stances" to interpret reality. Most notably, they have a physical-mechanical way of construing things as well as a psychological one that allows them to think about willful agents whose behaviors can be understood in terms of their beliefs and desires. Children also have distinct ways of thinking about biology, goal-seeking entities, and artifacts. All five of these ways of construing reality can be applied to a phenomenon, such as a person's actions. Each stance highlights different causal patterns, and they collectively provide a more objective sense of an entity. Children's early deployments of these interpretative stances set up the frameworks used by adults to navigate their gaps in understanding. Versions of infants' physical and psychological stances are now being proposed as built-in components of AI systems.

The recent decline of revealing mechanisms in children's lives provides an opportunity to better understand how all of us adapt to changes in explanatory resources and to shifting areas of understanding. Rather than neglecting and even actively refusing to think about gaps in our explanatory understandings, we can see how an awareness of those gaps creates an empowering form of intellectual humility. Sharing that humility with younger generations may be one of the most important ways we can help prepare them for the future. A major challenge to achieving humility, however, is a persistent tendency to think we have far more explanatory knowledge in our minds than we really do. From an early age we underestimate how much we rely on other people, and increasingly on artificial intelligences, to retain information and do much of our cognitive work—a trend that is only exacerbated by the decline of directly explainable things. We are becoming ever more outsourced and unaware.

Some of us are especially fascinated by devices and can't resist taking things apart to see how they work. Perhaps only this small minority will be affected by fewer childhood encounters with decipherable components. They may lose their drive to uncover underlying causal factors if that drive is never rewarded with insights, and this frustration may lead to a culture lacking a critical mass of such experts. More plausibly, the effects may spread far beyond people with passions for parts, including those who vehemently disavow any interest in technology or its underlying mechanisms. We may all benefit from having a better sense of the vanishing mechanisms and the possible effects of that loss on growing minds. Much will depend on how children learn to cope with different forms of ignorance and the attitudes they develop toward these different forms; adults will strongly influence those attitudes.

In this book, we will often compare the experiences of children and adolescents with those of their grandparents, using the terms *zoomers* and *boomers*. Zoomers have different information diets and preferences from their grandparents, who grew up in the fifties and early sixties. Their paths of discovery, patterns of outsourcing, and kinds of insights differ as well. Yet these differences have largely been overlooked. We will find ample evidence of changes in what they think about, which in turn leads to the more provocative question of whether there are also changes in how they think.

The first part of the book (Seeing Deeper) illustrates why insights and explanations are at the heart of higher-level human cognition. Chapter 1 opens with insight. In this book, *insight* is neither a sudden flash of understanding nor a complete explanation. It is more typically an increase in the ability to comprehend some meaningful part of a story. We then see how exploration, discovery, and insight are parts of virtuous knowledge cycles when embedded in the right kinds of real-world environments. These cycles in turn nourish the drive to wonder.

Insight and its virtuous cycles make clear why explanation and understanding are at the heart of a cognitively meaningful life. Chapter 2 examines explanations in more detail. A surprising variety of types of explanations infuse our daily lives; different types dominate thinking about psychological, biological, and physical processes. These contrasts help us appreciate the special status of mechanistic explanations in any analysis of how changes in what we are thinking about might also influence how we think.

The second part (Purposeful Parts) considers artifacts and their roles in evolving cultures and in growing minds. Chapter 3 documents the recent widespread and rapid transformation from easily interpretable parts of artifacts to causally opaque components. In doing so, it sets up the key question of how such a dramatic shift could influence the growth of explanatory practices and preferences in different generations. To see how artifact transformations can influence thought, chapter 4 describes other large-scale changes in artifacts at different points in time from prehistory to the nineteenth century. Apparent increases in the intelligence of individuals often arise from the evolution of cultures as repositories and transmitters of information, but artifacts are often neglected in such discussions despite being essential to the process. We focus on three antecedents to the shift to conceptually opaque parts: the rise, in prehistory, of devices with functionally heterogeneous subparts; the emergence of internally powered accurate clocks in the seventeenth century; and the rise of devices with interchangeable parts in the first half of the nineteenth century. These prior transitions have been described as having cognitive consequences, and childhood and adolescence are often described as where those consequences matter most. The recent shift from causal transparency to opacity is arguably much more momentous. In conjunction with an AI-fueled explosion of unexplainable information processing, the metamorphosis of devices' innards may have major consequences even as it has been largely ignored.

Chapter 4 then moves to the potential cognitive and societal consequences of this recent shift to a less explainable world. It first documents consequences for the contents of our thoughts, or at least what we talk about. In contexts ranging from natural language samples to advertisements for cars, a precipitous decline has occurred over the past few decades in talk about parts, their causal interactions, and, more generally, about the underlying mechanisms of everyday things. That shift and its link to the increased causal opacity of artifacts is startling in terms of its magnitude and pervasiveness. This link then raises the deeper question of how this change in content might influence the kinds of insights people seek as they grow up in the new world of digital technology.

Chapter 5 explores the origins of explanatory abilities in young children, considered initially as solitary explorers. Recent studies reveal that people, from infancy onward, have an interest in the hidden causes of surface patterns. Before they utter their first words, infants zero in on causal relations as opposed to mere correlations. They sense important things at work behind surface impressions. As preschoolers, they start to make sophisticated inferences about how some configurations of parts are related to mechanical complexity. We also see early preverbal indications of explanation-seeking curiosity.

Chapter 6 shows that, while intuitively appealing, construing cognitive development as solitary agents focusing on the here and now is not an accurate understanding of how children's minds grow. From birth, they are immersed in relations with other people. They develop skills for learning from others as well as abilities to think about other worlds and different points of view. A rapidly expanding research community has been uncovering the earliest components of the cognitive skills supporting these different ways of going beyond oneself. These skills grow and amplify explanatory understanding in several ways: navigating the divisions of cognitive labor that occur in all communities, evaluating one's own knowledge and that of others, thinking of other possible worlds, and creating innovations. The power of embedding our thoughts in larger networks of minds is often overlooked, as we mistake much of what we have outsourced as still being part of our own intuitive understandings.

Given the many ways that early explanatory abilities can grow during childhood, through both individual learning and from being embedded in social contexts, the final part of the book (Downstream) examines longer-term prospects. Most discussions of generational differences focus on adolescents and young adults to get a better sense of lifelong effects. Chapter 7 follows that lead and examines how the new digital world could have long-term influences on what people think about and on how they think.

Chapter 8 turns to how generational contrasts are often amplified when explanations and insights are immersed in social interactions. In tandem with a recent surge toward more nuanced rules of etiquette concerning diverse media communications between individuals and groups, other changes have occurred in how people distribute knowledge and thinking among members of a group. These patterns of information management can occur in real time and when people outsource explanatory components to other minds and devices. The ways people construe and navigate divisions of cognitive labor are also shifting.

Chapter 9 then explores the implications for our views of the explainable world. Radical changes are occurring in how we understand progress in science and technology. As AI systems are becoming less explainable while making ever-more-accurate predictions, the value of mechanisms and models is now being challenged. The idea of causation has been criticized as adding nothing beyond sophisticated analyses of correlational patterns. Yet some of the most sophisticated experts in artificial intelligence have persistent concerns about sacrificing our explanatory insights to advance predictive accuracy.

Chapter 10 takes stock of what we have learned and asks how we might best prepare children for the changing explanatory future and how those considerations can enhance our own lives in the present. In the end, we are confronted with both pragmatic and experiential concerns. The pragmatic concerns ask how current and future generations will handle challenges and advances to come in technologies and sciences. The experiential concerns ask what the human experience will be like if we surrender immediate agency and whole realms of insights in exchange for incremental advances in our physical lives. I suspect that many will debate the pragmatic concerns and that few will question the experiential ones. The debate may center around trade-offs between these two concerns and whether they are necessarily at odds with each other.

In more fanciful moments, I think of the replacement of explainable machine innards with alien components as a technological invasion analogous to the plot of one of my favorite movies as a child, *The Invasion of the Body Snatchers*. For the unfortunate few who don't know the film, an alien race exchanges new life forms with humans. Those new forms are superficially identical to specific people but have completely different insides, whose workings are mysterious. With today's artifacts, however, we are the aliens causing the change—a change that ultimately may make our minds more alien as well.

I Seeing Deeper

1 What's the Go o' That?

Many of history's greatest scientists describe childhood moments where a wonder-fueled exploration of an object led to a new insight and then to more questions.

As a three-year-old growing up in the mid-nineteenth century in Scotland, the future physical scientist James Clerk Maxwell showed a fascination with how things worked, often asking, "What's the go o' that?" He was constantly taking apart door locks, discovering hidden sources of streams, or tracing the pull wires that made house bells ring. When he found the answers, his mother described the young Maxwell as "a very happy man."[1] His affinity for mechanical processes inspired many of Maxwell's later breakthroughs in formal mathematical models, an influence he frequently noted.

The biologist Jennifer Doudna remembered insight-yielding events when growing up in a rural area near Hilo, Hawaii. She recalls realizing that certain plants folded when touched as a way of protecting themselves. Her insight generalized to the realization that different plants had distinct self-protection methods, such as thorns, noxious smells, and skin irritants. These insights led to more questions about the internal mechanisms that enabled protective properties. In a later interview on her research as a scientist, Doudna stated, "No one can take away those moments . . . when we saw something in nature that had never been seen before. You . . . get a little bit of knowledge about the world that unravels another piece of the mystery of nature. That's what I love about science."[2]

These experiences are not limited to a few gifted scientists. We've all had rushes of understanding and pleasure when a previously obscured component explained a mysterious pattern. A baby might be fascinated by a small box that rattles when shaken and then smile when she peers inside the box and discovers several metal balls. A kindergartner watches lawn mowers in their neighborhood cutting grass and wonders how they cut so quickly and evenly. When a neighbor hits an obstruction and turns a mower on its side to check the blade, the child has a surge of insight as to what cuts the grass. New questions may then arise, such as how the same mower can cut different lengths, leading to more insights.

Insights wouldn't happen without a drive to ask *why* and *how* questions and to seek answers to those questions. This drive can be described as wondering and, when nurtured and supported, it can lead to a life of exciting discoveries.[3] But wonder alone cannot yield insights. If the underlying causes remain hidden, an inquiring child may become frustrated and start to view wondering as futile. Achieving explanatory insights requires some degree of causal transparency. There must be ways to see beyond intriguing surface patterns and discover their underlying causes. None of the areas of fascination listed for the young Maxwell would be available to a young child today living in a modern urban home with electronic door locks, wireless controls, and remote signaling.

Characterizing Insight

We adopt here the Oxford English Dictionary's description of the early occurring senses of *insight* as "understanding into the inner character or hidden nature of things," a meaning that started to emerge in Middle English around 1175 and became commonly used in the 1600s to mean grasping what is beyond the surface. It involves learning why things are as they are, how they work, and how they got that way. Insight provides a glimpse into the causal infrastructure that maintains reality. It usually involves learning something new and different about an entity, not just an additional detail of existing knowledge. It resembles the sense of insight implicit in the adjective form *insightful*.

At the beginning of the twentieth century, the idea of suddenness became salient. In some classic psychology studies, *insight* came to mean abrupt breakthroughs in problem-solving and was described as "insight learning."[4] Another sense emerged of more deeply knowing one's own internal psychological states, a theme often appearing in the psychoanalytic literature.[5] Neither suddenness nor self-awareness are critical to insight as used here. We can view the *in* in insight as applying to learning more about the inner or deeper natures of all things at any pace that is productive. In most cases that deeper nature has causal implications of giving rise to surface-level phenomena.

Insight occurs in situations that provide interpretable clues to mechanisms and their components. Relevant parts of the world must be *explainable,* meaning that it must be possible to discover and understand a meaningful part of the mechanisms at work. In real life we rarely unpack the complete mechanism, but we learn enough to enrich our abilities to interpret, explain, and predict. We can also gain insight by learning from others, but there, too, success requires available and reliable sources.

Some of the greatest discoveries in history of science involved a long-term process with several detours and backtracking. These halting, nonlinear paths of progress appear in many scientists' notebooks, such as those kept by Michael Faraday and Charles Darwin. They often seem to tentatively entertain an idea for extended periods before fully committing themselves to the new insight.

Michael Faraday struggled to make sense of electromagnetism. Many of the phenomena he explored were only recently available for study because of the invention of reliable batteries. After Alessandro Volta discovered how to construct batteries in 1800, it was easy for people to make their own. These "voltaic piles" were stacks of alternating layers of two metals (usually zinc and copper) separated by paper saturated with salt water. Faraday used these batteries to conduct hundreds of experiments on the relations between electricity and magnetism. Yet even when the experimental evidence clearly pointed to a specific conclusion, he was cautious. For example, he once wrote to a friend, "I am busy just now on electromagnetism and think I have got hold of a good thing but can't say; it may be a weed instead of a fish that after all my labor I may pull up."[6]

Through page after page of Faraday's copious notebooks, his insights into the nature of electromagnetism emerged slowly and sometimes with pauses, tangents, and backtracking.[7] He often seemed to initially have a vague sense of a new pattern that gradually came into focus (figure 1.1). When he finally was sure, he certainly considered that he had an insight, but it was not a sudden revelation.

Similarly, Darwin did not at one moment in time suddenly say, "Aha! I see that species originate by evolution through natural selection." The idea became more and more dominant but yielded for a while to alternatives before gradually swinging back.[8] For example, he was repeatedly drawn to an

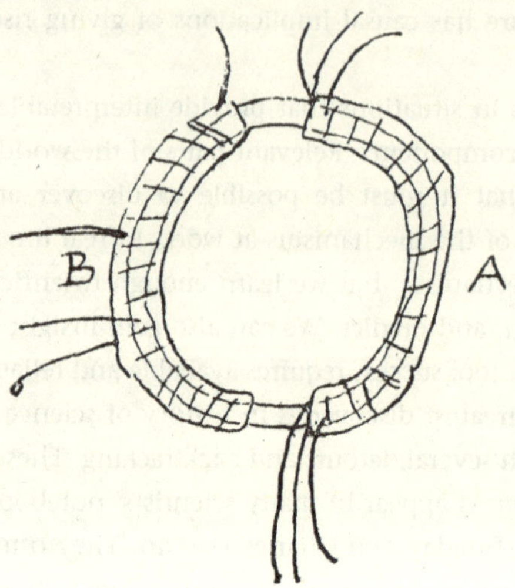

Figure 1.1
In Faraday's notebooks, his insights were gradual and hard-won, not sudden. When he drew diagrams, like this one of an inductive coil, he wrestled with how to interpret the coils in light of initially confusing experimental observations. *Source*: Thompson, S. P. (1898). *Michael Faraday: His Life and Work*. Cassel.

alternative reproductive model in which adaptations occurred in the course of an organism's development, a perspective not that different from Lamarck's.

We may retrospectively think that many insights are more sudden than they really were. Once we become firmly convinced of a new set of causal relations and realize how much they add to or clarify a previous position, we may distort what happened. The new perspective may seem so obvious and compelling that we may adjust our memory of the discovery into a more abrupt surge of understanding than it was. Later chapters discuss other forces that converge to support misleading impressions of insights as sudden breakthroughs in the minds of individuals.[9]

Insight is not normally a revolutionary conceptual change that overturns a prior theory or worldview. More often it is the act of uncovering a new class of causal relations that fits well with prior knowledge but expands on understanding in ways that go beyond simply adding more details. In other cases, the new realization may be the discovery of a surprising discontinuity from what is already known. When quantum physics rapidly unfolded at the beginning of the twentieth century, it was a radical departure from classical physics of the larger world of everyday experiences. Early studies like the double slit experiment, which showed how light could behave both as a wave and a particle, propelled physicists to dramatic new insights.

Once the causal processes responsible for surface phenomena have been revealed, it's not easy to undo that insight. If the finding is clear and well grounded in observations and empirical data, it is now part of how one thinks about the world. Most of the time that new understanding is a welcome addition. This can happen at any age. A few years ago, because of cataracts, I had both of my eye lenses replaced with artificial ones. I had not realized how monochromatic and blurry my world had become over the preceding decade. With the new lenses, objects had sparkling clarity and contained vibrant colors. It was a bit jarring at first and took some adjustment, but it was also inconceivable to think of returning to my cloudy natural lenses. When you come to see the world more clearly and with new characteristics, even if the process causes some initial discomfort, you rarely feel the desire to go back to a state of greater ignorance.

Sometimes the underlying structure may not be pleasant. You see how the sausage is made and would have rather not. Few people feel swelling gratitude upon first learning how the kopi luwak coffee they are enjoying is made. They are more likely to be shocked to discover that the coffee beans were fed to civets and then harvested from their feces. But even in those unappealing cases, many people, while groaning, would not choose to go back to ignorance. While some neuroscientists advocate treatments that can erase recent memories of trauma, it is the rare person who would want to give up an unpleasant insight about how a class of things is made or how an organism comes into being.[10]

How Insights Help

Insights may help make explanations more cognitively digestible by facilitating the transformation of an elaborate explanation into smaller conceptual chunks. Most natural phenomena and much of modern technology may require long lists of complex statements for state-of-the-art explanations, which still are likely to be incomplete. Insights enable us to expand our explanatory understandings and not be overwhelmed. They provide natural chapters of an explanation. When we realize that we now see why one phenomenon is different from another, we can stop and enjoy that insight and its cognitive consequences without having to dive deeper. We might initially think that octopuses and humans focus their eyes in the same way, but we have a new insight when we learn that, while humans focus on near and far objects by changing the shape of their eye lenses, octopuses focus by moving their lenses to different distances from their retinas, just as traditional cameras do. This insight brings with it a host of new questions. Why did the two species adopt such different solutions to bringing near and far objects into focus? What are the costs and benefits of each? What do other species do? These and many other questions offer attractive explorations, but the insight achieved also provides a natural end to one chapter and the opportunity to pause and reflect about other possible topics of interest. We may then decide to switch to an entirely new subject, or, instead, go more deeply into understanding octopus vision.

In other situations, an insight might show us how two things that we thought were completely different are the same at a deeper level. In such cases, we can also pause and enjoy the new idea and not feel the need to learn more details. We might learn that teeth, claws, and nails share a surprising number of structures and physiological functions and diverged from a common ancestral entity.[11] Here, too, new avenues of exploration open up, but we also sense a natural transition in our growing explanatory understanding that allows us to pause and decide whether to uncover more information or shift to a novel topic.

Insights often inspire a whole new set of what-ifs. Imagine learning that dinosaurs had their doom sealed when a large asteroid crashed into the tip of the Yucatan peninsula, creating an enormous debris cloud that resulted in climate change that was harmful for larger animals. Upon gaining that insight, it is almost impossible not to ask, What if the asteroid had just missed earth? Would there still be mostly dinosaurs and only small rodent-like mammals? What if the asteroid had landed somewhere else? By one set of calculations based on geology and vegetation patterns at that time, the asteroid would not have had such devastating effects if it had landed in most other parts of the earth's surface.[12] We now have a platform for launching many new lines of inquiry about alternate histories and their consequences—but we also see that we have reached a good place for stopping our line of inquiry given our other interests and demands on our time.

By highlighting coherent chapters of a complex explanation, insights can support the conversion of explanations into versions more suitable for long-term storage in human memory. To accommodate storage limitations, humans use various techniques for storing information more efficiently, but this process almost always involves distortions. This problem is not unique to humans or indeed even to organisms. It applies to any device with limitations on processing speed, storage capacity, or transmission, which currently includes all known systems. Accordingly, many devices employ compression algorithms. Information is compressed into a more efficient form for storage and transfer. In an ideal world, compression would be perfectly efficient and no information would be lost. In most real-world cases, costs do arise in what is known as *lossy compression*. Losses of this sort are common in human cognition because of the constant pressure to encode information in simpler terms. Insight may signal an opportunity to trim a representation into a more durable format that is easier to preserve. It may play a role analogous to clause structures in processing sentences, where the end of a clause prompts the recoding of its content into a simpler format as the listener transitions to a new clause.[13]

With insights and compression, we can now use the clusters of causal relations to explain the key parts of a system and how they work together. As an explanation about a complex system unfolds, insights help us detect the boundaries of functional parts, such as the magnetron in a microwave oven. Insights may highlight a transition to other components, such as the waveguide. When we see the transition, we abstract away from many of the details of the magnetron and store the key information in a form that is more durable and easier to store. In doing so, however, we tend to oversimplify and may cause distortions that can be problematic when trying to use the compressed versions later in design or diagnosis of a problem. On the positive side, insight-guided summaries may provide a version of the original causal pattern that is more generalizable because it is more abstract and applicable to a wider range of instances.

We can now see why the development of insight in children bears on questions about insight at all ages. Some early emerging forms of explanations may play outsized roles in fostering insights. Mechanistic explanations are a notable example. Changes in exposure to mechanisms may therefore be particularly influential. Areas where insights are easy shape the realms of the explainable. Some kinds of explanatory structures fit especially well with certain stable patterns in the world. When a child hears references to psychological processes in physical terms such as greasing the skids, being crushed, achieving a breakthrough, getting into a groove, or having leverage, their understanding of the psychological processes might be enriched by having a well-developed set of physical and mechanical intuitions. Their understanding might also be distorted by trying to fit the psychological too tightly into the mechanical. Thus, they might assume some processes are more discrete and deterministic than they are, or they might imagine the pace of change as faster than it is. We explore these issues in later chapters.

How will ordinary understandings of the world around us be different when today's children, and their children, are adults? Some kinds of shifts in insight and understanding seem likely. Future generations may regard clockwork mechanisms as quaint curiosities in relation to their more enlightened views of digital signaling and information trails. They might regard questions about underlying mechanisms as unwelcome distractions from the more important tasks of learning about functions and knowing how to operate user interfaces. Whatever futures we encounter, we have more influence over our cognitive destinies than we realize. Instead of seeing ourselves as drifting helplessly in turbulent rapids with ever-larger unexplainable barriers, we can be active voyagers, navigating the flow of causal information with interest and insight. We may not be able to go upstream, but we can have considerable influence on where we end up and how we get there. Neglect of changing explanatory abilities may limit our attempts to help future generations prepare for what is coming. Because we created this new world, we should provide the best assistance possible to those that follow.

Insights, Motivation, and Exploration

Insights are rewarding. When we wonder, explore, and discover things, insights are associated with the pleasure of understanding something more deeply and with making progress in building a larger-scale explanation. When those same activities do not end with an insight, eventually we disengage and focus our efforts elsewhere. Children and adults have a robust intrinsic curiosity that can persevere, and even increase for a time, when no insights are immediately found. However, when curiosity does not yield new insights over an extended period, we tend to redirect our efforts.[14] Rewards may also unfold over time. A new insight can allow us to have future experiences with greater pleasure. A few years ago I learned that, when dogs come across a new odor, they don't just remember the location where they encountered the odor; they also note and remember the changing gradient of odor intensity over space and remember the direction an animal was moving.[15] Ever since then, watching our dog racing about absorbing new smells has been a completely different and more rewarding experience for me as I now envision all the cognitive maps and vectors she is constructing of animals' movements.

Insights are typically embedded in wonder-driven PHED cycles that consist of puzzling, hypothesizing, exploring, and discovering.[16] When I first described children as engaging in such cycles, I did not consider how discoveries can be demotivating rather than the rewarding. Yet this can easily happen when one discovers a hidden component that stubbornly fails to yield new insights. Imagine a four-year-old who became interested in why one toy moved in a serpentine path and another went straight. In the 1960s a child might have hypothesized that something inside was making the difference and, after figuring out a way to take it apart, discovered how an oval-shaped disk in one created serpentine paths while the circular disks in the other propelled it in a straight line. That boomer child gained a revealing insight that would carry forward to other mechanical systems. In contrast, a zoomer child might discover the relevant internal parts to be indistinguishable electronic circuits. The zoomer child's discovery would likely be a disappointment in contrast to the boomer child's. The mere discovery of the parts responsible for a surface phenomenon may not be rewarding if those parts fail to reveal new insights.

Perhaps we shouldn't worry about fewer insightful encounters with causally interpretable components and their possible motivational consequences. After all, predictions of technology-induced cognitive deficits have popped up like weeds over centuries and seem inevitably to be false alarms. Elders have belittled younger generations as less intelligent because of cultural changes, but their consistently mistaken pronouncements often seem silly in retrospect. Socrates saw the rise of the written word as hindering the memory skills required by oral transmission of cultural information.[17] With the surge of printed books after Gutenberg, easily available books and documents were predicted to become cognitive crutches that would make mental muscles wither.[18] Similar claims arose with the emergence of radios, televisions, and personal calculators. In every case, predictions of cognitive decline were wrong. In addition, we may have a cognitive bias to discount the value of new technologies that we believe emerged in our lifetimes rather than being present before we were born.[19] Despite this record of pessimistic mispredictions, displacements of old technologies by new ones do not always enhance the mind without costs.

Historically, new technologies have made some workplaces worse for many workers. The growth of assembly lines in manufacturing plants in the twentieth century reduced the jobs of line workers to repetitive and boring sequences of actions. Those workers, or their offspring, were predicted to swap those meaningless jobs for more interesting ones as manufacturing became more automated, but that outcome has not always happened.[20] We are also witnessing a pace of change that is unprecedented. Moore's law, proposed in 1975, asserting that the number of transistors in a circuit of a given size would double every two years, seemed fanciful and clearly wrong—but after fifty years, while its future is more hotly debated, it is still endorsed by some leading chip experts.[21] The surge in circuit complexity helped launch a recent leap forward in AI, which has alarmed several leading experts. We don't have to be doomsayers to appreciate how technological progress can also include pitfalls.

The negative consequences of growing ignorance of causal components might also seem concerning if one takes the view that only a small proportion of the general population, say five percent, needs to know anything about mechanisms. But that small percent would still founder growing up in a world of undecipherable things. How is their especially active wondering about mechanisms to survive if they never have rewarding insights? Those who profess no interest in mechanisms should still want to ensure that the critical minority has the experiences it needs.

It is implausible, however, that disappearing mechanisms are a problem for just a few members of a society. Perhaps most people need a shared body of mechanical intuitions to serve as a cognitive framework that enables communication and problem-solving in groups. This proposal can be illustrated with a story retold many times in my family. For several years, my mother drove a lime green Fiat station wagon. While it was undeniably cute in comparison to the lumbering American station wagons of that time, it was a maintenance nightmare, constantly in the shop for repairs. The bills from the nearby Foreign Car Repair Experts shop were typically two to three times the bills for comparable repairs on our Ford sedan. Even though my mother had long claimed that she knew nothing about how things worked, she became suspicious of the repair shop's charges and decided to investigate.

When she announced her decision at dinner, my father looked at me with a grin and said, "Hell hath no fury like your mother being overcharged."

The next day, my mother started interrogating the shop manager about the charges. According to her recounting of the conversation, the manager first replied that the parts cost much more and it was much more difficult to work on our Fiat. But when my mother asked to see the parts prices, they were only a little more than those for our other cars. The manager then claimed that anyone who really understood cars knew that the Fiat wagon was much more complicated to work on. My mother then asked him to explain why it was more complicated. She asked how many parts were involved in each one of several different repairs ranging from replacing the clutch to servicing the brakes. There was no obvious difference from comparable repairs on our Ford. She asked him to show why it was more difficult to remove and replace defective parts. He had no answer. The deeper she drilled into the details, the more he became flustered and eventually tried to blame one of his employees for the excessive charges. My mother left that repair shop and never returned. She found a less expensive and better-quality repair shop in a nearby town.

This story reveals a subtle level of mechanical intuitions in someone who professed to have none. By simply growing up and overhearing talk about various car parts and different kinds of repairs over the years, she had built up an implicit sense of the approximate numbers of parts mentioned in certain repair jobs, how long different repairs took, and what it meant for a repair job to be difficult. She knew what mechanical questions to ask the manager and what responses sounded fishy. She knew what kinds of evidence were relevant. Young adults today may not have had the opportunity to develop that tacit knowledge shared by most adults sixty years ago.

Vanishing Explanations

We will examine in detail the recent shift from explainable parts to inscrutable electronic chips. Our examination enables us to better understand how the disappearance of a cognitively comfortable and early emerging explanatory realm might change the language and thought of future generations. This includes the ways we understand and manage our growing areas of explanatory ignorance. Children are especially informative in this regard given their traditional abilities to learn and cognitively prosper despite having much larger explanatory gaps.

I suggest that the conceptual changes associated with artifacts may have a special importance with implications that extend far beyond how we create and use tools and devices. As that account unfolded as a central theme of this book, it began to resonate with the broader and even more recent surge in the unexplainable in almost all areas of human inquiry. Across the sciences, revolutionary advances in machine learning and AI have provided remarkable improvements in predictions, at the cost of occurring through indecipherable computational processes. This transformation raises the problem of how to navigate a world where traditional expectations of understanding and explanation are no longer realistic.

The successes of these new computational techniques might be taken as showing that we no longer need to understand how things work. In the natural sciences and medicine, a reliance on brute-force calculations and data-driven predictions can displace mental models and mechanistic understandings. These computational processes, alien to human cognition, deliver results that may enhance predictions but lack an intuitive connection to how we have perceived and explained the world throughout history. We have relied on mechanistic models to simplify phenomena by isolating a few key variables and their interactions so that we can understand, manipulate, and innovate. Such models and theories invariably distort reality to some degree, but the distortions introduced by processes such as idealizations and abstractions were acceptable losses given the gains achieved as the sciences and technologies advanced. However, it now seems that continued progress in many areas may defy such simplifications, requiring us to abandon explanatory models in favor of purely predictive approaches. How should we adapt to this new reality? One approach might be to resign ourselves to the same role that the repair shop tried to assign to my mother: we are not cognitively equipped to understand the basis for the foreign AI's recommendations and should accept our limitations and embrace what those superior experts tell us. Alternatively, we can all have enough understanding and outsourcing skills to be able to evaluate and work with these new artificial systems as intellectual collaborators and learners. This is how young children learn from more knowledgeable adults. We will initially explore these issues with artifacts and then turn to a broader examination of the ways we might continue to grow as individuals and as a culture given the evolving nature of explanation and understanding in our new technological and computational world.

* * *

We all seek new explanatory insights from an early age, and traditionally in our quest for explanations we have found many everyday artifacts to offer ample opportunities for new insights. But contemporary devices, tools, and appliances may no longer afford such opportunities directly and thereby thwart the growth of an entire realm of explanatory understanding. This may be critically different from more localized technological shifts and bears on broader changes in the bounds of explainability. To more fully appreciate the scope of this transformation, we must first survey the surprising variety of types of explanations.

2 Explanations and Mechanisms

Few animals are tougher than tardigrades, also known as water bears or moss piglets (figure 2.1). They have been found thriving on all continents, including Antarctica, and have remained largely unchanged for over five hundred million years. Over one hundred thousand tardigrades might be found in a single gallon of water taken from either a freshwater pond or a marine swamp. They are barely visible—about the size of a small grain of salt—and most people are unaware that these tiny creatures are all around them. Tardigrades can survive in conditions that are lethal for most other organisms, including extremes of temperature, air pressure, and malnutrition. They may be most impressive in their abilities to survive radiation exposures over a thousand times the lethal dose for humans. They were found frolicking in radioactive puddles near the center of the Chernobyl disaster.

Upon learning of this ability to survive massive levels of radiation, one may immediately want explanations of how this is possible. What is it about their physiology and anatomy that makes them so different from other animals? While several factors may be involved, a recent discovery describes an extraordinary ability to repair damaged DNA.[1] Tardigrades are not immune from the ways that ionizing radiation damages genetic material. They differ, however, in their ability to quickly ramp up the production of genes for repairing DNA, and they can accomplish these repairs rapidly enough to survive extended severe radiation exposures.

This brief example illustrates many aspects of explanations. We see how a novel phenomenon automatically poses explanation-seeking questions of interest to people far beyond the members of relevant research groups. We sense the dawning insight upon learning about their special ability to repair DNA. We feel that we have received the right kind of explanation for the question at hand, namely one that describes internal molecular causal mechanisms that provide a coherent account of how these creatures can survive when most other animals do not.

We also soon realize that our insight has launched new questions requiring explanations and revealed new, previously unappreciated realms of the unknown. For example, why are tardigrades different from most other species in their ability to survive radiation? Are their abilities to survive other kinds of environmental assaults, such as extreme temperatures and altitudes, related to the same underlying mechanisms? How does the ramping up of the DNA repair genes happen? What triggers the change, and how is it regulated? Could this enhanced repair process be adapted to fight mutation-related diseases in humans? Answers to each of these questions will almost always launch a new wave of questions.

The tardigrades example also illustrates how, after learning secondhand just a few sentences about a completely novel organism, adults immediately seek explanations of unusual properties. These, in turn, lead to follow-up lines of inquiry. This sophisticated combination of exploration and inference has roots in infants' firsthand encounters with objects, many of which are mechanical devices. Those early experiences provide versatile models

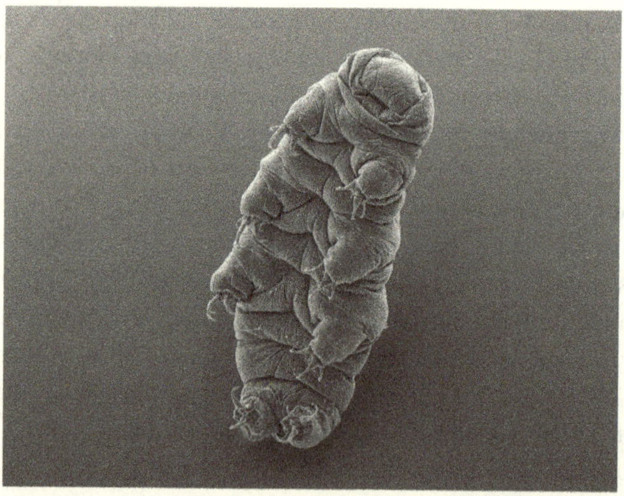

Figure 2.1
Tardigrades illustrate how learning about an unusual trait in an unfamiliar animal can immediately trigger a powerful desire to ask questions that ultimately lead to a new cascade of explanations. *Source*: Image courtesy of Bob Goldstein and Vicky Madden, University of North Carolina.

of how underlying causal explanations can provide powerful insights that make life more interesting and meaningful.

Kinds of Explanations

Depending on our goals and the phenomena we are encountering, we prefer certain kinds of explanations over others. Using a different example, consider in more detail the various ways we can provide explanations about coyotes. We can explain how a particular coyote, let's call it Wile E., came into being and grew up to be the current Wile E.: an ontogenetic explanation of an individual. We can explain how coyotes in general develop from conception to maturity: an ontogenetic explanation of a kind. We can explain how coyotes as a kind emerged over the course of evolution. We can explain how the parts of coyotes work together to enable coyotes to survive as stable entities over time: a constitutive explanation. We could also give a constitutive explanation for an individual coyote, but that would be oddly redundant if the same explanation applied to all coyotes.

Explanations can also be mechanistic or functional. Mechanistic explanations describe how an entity works in terms of its parts while functional explanations describe the roles of various units in an entity. I explain how my car works functionally by describing how putting the key fob in its holder turns on the car, how one lever puts it in drive and reverse, and another lever controls the lights. I explain how the steering wheel makes it turn and how the brake and accelerator pedals make it slow down and speed up. I can put together a much more elaborate functional explanation of this sort without having any idea of the causal mechanisms underlying such processes as braking and steering. In modern cars, some car parts with identical functions can work differently at the mechanistic level.

It is rare to have a mechanistic explanation for a thing but not a functional explanation. If I know how a car's brakes work in terms of the mechanical interactions of its parts, it is difficult to not know the functions of those parts and of the brakes as a whole. A notable exception occurs in reverse engineering challenges, where an artifact's mechanisms are clear, but its function is unknown. A classic example is the Antikythera mechanism, a two-thousand-year-old device discovered in a shipwreck off the Greek island of Antikythera in 1900. Over the succeeding decades, researchers were able to reconstruct all the working parts of the corroded bronze object and could fully explain the mechanical operations performed by its gears, wheels, and cranks. The Yale physicist and science historian Derek Price described a reconstruction of the entire mechanism in 1974 and proposed that it was a device for computing astronomical patterns, a function that finally is achieving consensus more than a century after the object's discovery.[2] Devices designed to predict patterns are where many reverse engineering challenges are found.

Mechanistic explanations have a privileged status in science and engineering and are where changes in artifacts' innards are likely to have the strongest cognitive consequences. We will return to a more detailed discussion of mechanistic explanations and their cognitive characteristics later in this chapter. That discussion is informed by first covering some more general features of explanations and how they are used.

While much of our focus in the following chapters is on explanatory insights concerning artifacts, we need to keep in mind how we think about causal explanations for all phenomena. The contrasts and similarities between such things as animals and tools can be deeply informative about both. Most of us would seem to gain insights into organisms primarily through information provided by others, not by taking them apart and deciphering their causal underpinnings. But insights into the workings of devices may frame how we try to explain aspects of organisms. Those mechanical intuitions heavily influenced the rise of molecular biology in the second half of the twentieth century.

We will not consider explanations in mathematics or formal logic, where a proof might be "explained" without any appeals to causality. As used here, a good explanation provides insights into a network of enduring causal relations responsible for a phenomenon of interest. These kinds of explanations are especially important to us. Their causal relations are also at the heart of decisions to act on the world. If you are trying to decide whether your broken washing machine is worth fixing, your decision is greatly informed by having at least a schematic sense of the primary components of washing machines and their own patterns of internal complexity. If your washing machine fails because a small sensor switch became clogged, you will likely make a different decision than if the primary motor burns out. With a rough sense of the components and how they work, you can make more informed decisions about what actions to take.

Useful explanations increase understanding of phenomena in ways that traditionally also lead to more accurate predictions and, most importantly, to insight. Yet, even as they provide insight, explanations are typically incomplete. They provide explanations for previously mysterious patterns while also uncovering new questions and unknowns awaiting further exploration and discovery. Upon learning that bioluminescence occurs only in saltwater organisms because it depends on two chemicals that can work only in salt water (luciferin and luciferase), one can simultaneously feel a surge of insight and think of a host of new questions as to why those two chemicals can't work in fresh water and why alternative forms of bioluminescence haven't evolved for freshwater organisms.

Finally, explanations are usually communicative in nature. Explanations are offered to others to provide insight into a question or puzzle. We can provide explanations in real time or across time intervals through various ways of storing and recovering information. Explanations can also be nonverbal. We have all encountered illustrations that provided compelling explanations. While their communicative nature suggests that explanations require at least two individuals, they can occur in one person through *self-explanations*. These occur when a person attempts to explain something to themselves. Several studies demonstrate that people can gain valuable insights by self-explaining, which can be understood as a form of internal conversation between different realms of ideas and models.[3]

Evaluating Explanations

Some specific explanations are obviously worse than others. Humans of all ages often must choose between competing explanations based on some metric of quality. Evaluations of explanations typically occur along two dimensions: beauty (explanatory aesthetics and preferences) and fits (how well types of explanations fit with specific sectors of reality).

Beauty

An entity is *explainable* in our sense if we can identify its critical components and how they interact to produce the properties of the entity as a whole. Good explanations honor a combination of clarity, elegance, and comprehensiveness. These attributes have sometimes been equated with beautiful explanations, or what amount to an aesthetics of explanation. Einstein was convinced that beauty was central to good explanations in physics and that "ugly" explanations, no matter how accurate, were insufficient.[4] In chapter 5, we will see how young children can also appreciate the beauty of an explanation.

Coherence and consistency indicate good explanations, yet people often believe coexisting contradictory or inconsistent explanations. For example, while many people will explain the seasons by explaining that the earth is further away from the sun in the winter, they also believe that it is summer in Australia when it is winter in North America. Only when the two beliefs are highlighted side by side do they suddenly realize that they cannot coherently believe both things.[5]

All other things being equal, people prefer simpler explanations. But people also use complexity as a cue to how fully an explanation is likely to capture what is going on. We are wary of brief explanations as oversimplifications that miss critical nuances. We usually consider both simplicity and complexity in trying to find the balance between the two, looking for the simplest accounts that capture the phenomenon most completely.[6] People prefer simpler explanations when no mechanisms are supplied in an explanation but more complex explanations when mechanisms are supplied.[7]

Breadth, the range of phenomena associated with an explanatory goal, enhances explanations.[8] Similarly, a diversity of evidence supporting an explanation makes it more appealing. A single account that explains how both air conditioners and dehumidifiers work is preferred to two different explanations. But breadth shouldn't come at the cost of losing insights into each of the entities involved. You might explain all the properties of animals by saying those properties exist because they were selected for in evolution, but that is a deeply unsatisfying explanation because it doesn't advance an understanding of any given animal. Simply saying that camels have humps and kangaroos have pouches because those properties helped them to survive doesn't give us new insights about those adaptations.

The order in which information is presented affects perceived explanation quality. For example, people prefer explanations in which functional information precedes mechanistic information.[9] In one study in our lab led by Mandy McCarthy, people were shown an image of a novel machine and asked in what order a teacher should explain the machine: first describing what the machine is for and then describing how the parts of the machine work, or first describing how the parts of the machine work and then describing what the machine is for. The function-first version was strongly preferred. It is not yet known why this preference is so strong.[10]

Fits and Misfits

Explanations come in different kinds that each align especially well with specific slices of the world, or content domains, but those alignments are never perfect. They typically create distortions and omissions of information even as they also provide cognitive benefits. Judgments of good and not-so-good fits involve assessing the costs and benefits of applying different explanations to phenomena in a particular domain.

Consider again origin-based versus constitutive explanations of organisms, machines, and nonliving natural phenomena such as geysers. Origins accounts can be either about the history of a single entity or about the history of an entire class of things. Thus an explanation can describe how a particular car was made from raw materials or how the class of things known as cars came into being. Similarly, an explanation can describe how a particular dog grew into its adult state, beginning with the moment of conception, or how the species of dogs came to exist through evolution. A scientific explanation of the origins of species will rely heavily on probabilistic patterns and population distributions, while the story of a single dog may stress a more linear chain of events. An account of the origins of cars would have a different character, presumably focusing heavily on goals and intentions of the car makers. These fits between explanation types, kinds, and individuals all can change over time, as well. Before Darwin, intention was more central to origins accounts of species.

We may use an interpretative format such as a mechanistic framework to facilitate explanation generation and understanding in a specific domain. Those formats rarely fit perfectly and can therefore skew interpretations, such as construing some biological processes in excessively machinelike terms. For example, some biological phenomena are best understood as a complex set of signaling pathways and feedback loops whose operations can change dramatically in response to changes in the body's state and the outside environment. These kinds of processes do not usually conform easily to canonically causal mechanical chains.[11]

Interpretative Biases

We think of explanatory fits as the ways in which a class of explanations handles some aspects of a phenomenon well but underplays or ignores others. Two explanations may highlight different real properties and relations, or one may fit so badly that it isn't even considered. In contrast, interpretative biases occur when a specific distortionary influence occurs across several domains.

Essentialism

Humans gravitate toward explanations invoking essences—that is, toward the idea of fixed, usually internal, properties responsible for a phenomenon of interest. Adults find essence explanations especially attractive for organisms (more than for artifacts). We might want to explain stripes on a zebra in terms of an internal essence, such as DNA, that causes those stripes to appear, but we wouldn't explain stripes on a jacket in the same way. Most distinctive patterns on things like clothing and furniture arise from the intentions and actions of external agents and not by factors physically inside those things. The immediately compelling nature of such differences led to studies on the early emergence of such intuitions in children starting in the 1980s.[12]

Explanations of why zebras have stripes are appealing when they refer to something intrinsic to zebras, such as tiny inside properties that cause the presence of the stripes. Essentialism, however, is often used to explain properties that, contrary to intuitions, do not arise from fixed essences. These include psychological traits that are attributed to groups because of beliefs about nonexistent essences associated with categories such as gender or race. Unfortunately, the explanations that we find most appealing are not just wrong; they can also be used to justify differential treatments of specific groups of people. But the presence of such biases may also confer benefits, especially early in development, as we will see in chapter 5. Here, it suffices to summarize some patterns in adults.

By focusing on the serious problems introduced by essentialism, we can overlook its heuristic value. Essentialism invites people to go beyond the obvious and look deeper, a theme repeatedly emphasized by the leading researcher on essentialism, Susan Gelman.[13] In many instances, we benefit by looking for underlying causes. The assumption of an underlying fixed essence may be an especially easy way to initiate and maintain a strategy of looking deeper. It may be easier to visualize than the more accurate idea of relatively stable, mutually reinforcing, causal property clusters.

Gelman has suggested essentialism may lead to beliefs in microstructural properties that are invisible to the naked eye.[14] Such beliefs may be useful given that much of modern biology consists of explaining how surface properties and behaviors arise from smaller—often microscopic—processes inside an organism. However, a misstep occurs when we assume those inner processes must be essential and fixed. That last assumption may be appealing because it is easy to state economically and to keep in mind, but it leads to conceptual errors. The philosopher Sarah Jane Leslie has maintained that essentialism is almost never true, even for chemical elements.[15] For organisms, laypeople often think of specific DNA sequence as a species' essence. However, no unique string of nucleotides corresponds to tigers' DNA. Instead, tigers represent a distribution of closely related DNA sequences. Without such distributions, evolution through natural selection could not occur. *Selection* means choosing candidates that differ from each other.

Reductionism

A different but related bias is a fondness for some kinds of reductionist explanations, or explanations that refer to a lower level of explanation as found in the traditional disciplinary chain: psychology reduced to biology, biology reduced to chemistry, and chemistry reduced to physics. The reduction from psychology to biology is especially seductive and can lead people to embrace explanations that are empty but invoke neuroscience in "explaining" a psychological phenomenon. Deena Weisberg and I had a strong intuition along these lines and, with help from others, tested it by creating passages about supposed findings that were vacuous but either did or did not shift between levels.[16] For example, the following two explanations were offered (to different people) about the curse of knowledge effect, a finding where we are more accurate judging our own knowledge than knowledge in other minds:

> The researchers claim that this "curse" happens because subjects make more mistakes when they have to judge the knowledge of others. People are much better at judging what they themselves know.

versus

> Brain scans indicate that this "curse" happens because of the frontal lobe brain circuitry known to be involved in self-knowledge. Subjects make more mistakes

when they have to judge the knowledge of others. People are much better at judging what they themselves know.

People realized that the first explanation was circular and empty but were seduced by the neuroscientific language in the second and thought they had gained explanatory insight when they hadn't.

Explanation Inflation

How well do we know what we can and cannot explain and what others in broader knowledge communities understand? We tend to think people have greater explanatory understanding than they really do, and people have especially inflated estimates of their own explanatory understanding.[17] We may be confident that we can provide detailed explanations of how everyday things such as zippers, staplers, and toilets work, only to be surprised at our ignorance when we are asked to provide such explanations.

Leon Rozenblit and I labeled this misestimation the *illusion of explanatory depth*, or the IOED. Over the past two decades, the IOED has repeatedly been found across cultures and in both children and adults.[18] The illusion is especially strong for estimates of explanatory knowledge as opposed to estimates of how well one knows facts (such as the capitals of countries), procedures (how to make an international phone call), or plots of movies (what happened in Casablanca). Even regular bike riders, when asked to identify a drawing of a mechanically correct bicycle, often pick drawings with erroneous depictions of the basic elements of a working bike, choosing a bike that would be physically impossible to ride.[19]

Despite the IOED, people can still retain high-level causal impressions of how an entity works. These might include knowledge about its energy properties (e.g., whether it is energy consuming, energy producing, or neutral); whether it operates fully autonomously or needs external control; how mechanistically complex it is relative to other entities; and whether it is wholly mechanical, a hybrid of mechanical and electronic parts, or wholly electronic. People may also sense whether a thing predominately engages in separating, mixing, crushing, or cutting.[20]

These gists may not all fit together neatly even at a high level. Major gaps can persist, but these patchy composites can still enable us to understand how a device works when it is available for inspection and tinkering. Thus, most people can easily figure out how a manual stapler works when it is in front of them even if they cannot produce a full working explanation from what they have stored in their memory. They can also use these bits of knowledge to evaluate the credibility of sources and to know what sorts of people are likely to be relevant experts. Indeed, the IOED can be construed not so much as an illusion of knowing in general, but as a misplaced sense of the locus of knowledge. It may signal a fairly accurate assessment of the relative level of explanatory understanding in a knowledge community, just not in individual minds. It underestimates how much we off-load explanations to other sources. Thus, the IOED influences how we manage, or fail to manage, our various forms of ignorance. It also bears on how well we understand what lies behind recommendations made by artificial intelligence systems: users tend to overestimate their abilities to explain how an AI made a recommendation or prediction.[21]

Adaptive Strategies

When we cognitively engage with explanations, we usually cannot fully process all aspects of those explanations and their implications. We need to adapt in ways that incorporate the limitations of memory, attention, and processing. We frequently employ adaptive strategies to reduce the load. These strategies rely on idealizations and abstractions, layers and partitions, outsourcing information, and revealing counterfactuals.

Idealizations and Abstractions

In all attempts to make a science more lawlike, researchers introduce idealizations, such as frictionless planes in physics or idealized populations in biology.[22] *Abstractions* are distinct from idealizations. Abstractions omit details (sometimes unwittingly), while idealizations intentionally change a detail away from what is true in the world so that a model works much more cleanly. More informally, we all simplify complex real-world situations by omitting details or by changing messy details into cleaner, more streamlined versions. Almost any attempt to explain real-world phenomena requires some amount of abstraction and idealization. If you want to understand how robins build their nests, you'll want to talk about nest materials at some level of abstraction. You may refer to twigs, grasses, and mud as components, but ignore many specifics of each. You may also describe typical shapes in idealized terms, saying that robins' nests are circular while knowing that they aren't perfect circles. As adults, we idealize and abstract automatically all the time, yet our ability to do this is not well understood.

Layers and Partitions

How does a refrigerator work? How does mitosis happen? Ask an individual for full explanations and they will almost invariably be unable to provide every element. In some cases, a full account may be known in a community, but it may exceed the cognitive capacities of any individual mind. Or perhaps everything can be known by one mind, but only after years of intense study and at the expense of learning much else. In other areas, the full details may not yet be known. Scientists don't yet fully understand aspects of mitosis such as the causal intricacies underlying metaphase alignment, the elegant organization of a cell's genetic material into neat rows of chromosomes and their partners on either side of a metaphase plate.[23] Even in the case of artifacts, we may rely on certain patterns that we do not fully understand, such as the condensation processes of the refrigerants we use in refrigerators.[24] How, then, do we know when we have enough knowledge for our present and future needs?

We rarely seek out exhaustive explanations that explain causal factors at every possible level. When we want an explanation about how human hearts work, we are usually satisfied if the initial explanation discusses the causal interactions between units at a level corresponding roughly to the heart's four main compartments and the valves between each compartment. However, explanatory goals matter. If we need an explanation of how the heart's contraction smoothly moves blood through those chambers, we need to know more about the intricate sequences of muscle contractions across the heart. This part of the story is nontrivial, as asynchrony between the ventricles and atrium can halt blood flow and be fatal, yet this level of detail is normally omitted in initial explanations of how hearts work.

We benefit most from explanations when we sense the layered, coherent organization of many natural and artificial phenomena and see how explanations are also layered in ways that capture these real-world causal patterns. Many natural and artificial entities have recurring, locally isolable elements. A sense of layering and stable subunits help manage the flood of information that accompanies exhaustive explanations. We can gain useful insights while also having large gaps in our own understanding. For example, if you want to learn more about how a classic train's locomotive steam engine works, you might be seeking an explanation at the level of units such as the firebox, boiler, valves, cylinders, pistons, and cranks and the interactions between them. The explanation itself would involve a mix of functional information (e.g., the firebox heats up water in the boiler) with mechanistic information explaining how the function is performed (e.g., An easily replenished firebox containing burning coals heats up air in pipes that run through an enclosed metal boiler full of water. The hot air pipes increase the water temperature until it turns into steam, which develops a sufficiently high pressure to power the locomotive.). In some cases, you might want to dig deeper and understand subparts of a boiler and their functions, but this is not necessary much of the time. Most people do not want to dive into the thermodynamics of steam under pressure even if it is ultimately critical to a full explanation of steam engines.

Outsourcing

All cultures develop by apportioning work into different categories with experts in different areas. This leads to divisions of cognitive labor—but simply mentioning divisions of cognitive labor glosses over the complex processes of determining who knows what and learning how to evaluate the quality of their knowledge. Over the last two decades, studies have been uncovering intricate skill sets in adults that start to develop in early childhood.[25] But these skills are not foolproof, and their development can be stunted if they are undervalued and their growth is not supported. Providing good support in turn requires an awareness of explanatory understandings and gaps both in one's own mind and in the minds of others.

Ignorance can be embarrassing, and it can be painful to discover a hole in one's knowledge. Such reactions and emotions may be appropriate in cases where one absolutely should know what's going on, such as during a revered family ritual. Much of the time, however, the discovery of one's own ignorance should be a source of pleasure and interest, as it provides a new opportunity to explore and learn. Consider the analogous case in the exploration of a physical environment. Learning about the existence of a previously unknown hidden area in a beloved park is often accompanied by a rush of excitement and a joyous sense of adventure. The same reaction should be at least as positive an experience when one is uncovering whole new terrains of intellectual adventures.

We often address the problem of not being able to know everything by knowing how to retrieve knowledge from others. This retrievability may lead to attributing external knowledge to knowledge that we have in our own heads.[26] Such misattributions can be problematic when we cannot easily access the appropriate expert. This kind of knowledge misattribution can occur in lifelong couples who reflexively cover each other's knowledge gaps without knowing they are engaging in those actions.[27] For example, it has taken me decades to realize just how little I know about cooking, as my ever-patient life partner graciously and quietly fills in my many lapses in even the simplest acts of food preparation. These forms of complementary cognition are also common in professional settings, such as airplane cockpits and ship bridge decks, where groups of people have been working together for years.[28] The effects also occur in families, where children are unaware of how much their daily cognitions rely on subtle cognitive scaffolding by their parents.[29] In many situations this misplaced sense of knowing may be adaptive; it helps restrain us from diving too deep into details and thereby become overwhelmed by an explanatory flood.[30]

We need to pace ourselves as we expand our understanding. We must allow partial understandings to stabilize and consolidate with other bits of knowledge. Upon attaining one insight, we often need to surface and take a metaphorical breath before diving deeper to gather new insights. Gaining insight is immensely rewarding, and insights can endure for a lifetime, but like so many skills and physical conditioning, insight should be appropriately paced. We may use insights to mark meaningful causal boundaries of complex explanations in a manner analogous to how we compartmentalize scene knowledge as we walk through a physical space.[31] We parse our journey into a series of episodes where the boundary of each episode prompts us to reformat the information from the preceding episode into a briefer, more durable format. Feelings of insight may serve the same role when we encounter a complex explanation.

Considering Counterfactuals

Counterfactuals have long been regarded as central to explanations and insight. David Hume described their power as revealers of *difference making*, the idea that a property or relation was causally important if its presence or absence made a difference.[32] In the last few decades, scholars across the cognitive sciences have examined how thinking about possible worlds provides new insights.[33] At first, creating and using counterfactuals seems straightforward. Simply change a true situation into a false one and consider the consequences. However, unpacking that process reveals a complex decision that is dependent on explanatory goals, the target of the explanation, background knowledge, and other factors.

Counterfactuals enhance learning about how things work or why they have certain properties. When trying to learn how a flush toilet works, you might pose counterfactuals to guide your investigations. For example, if you ask what would happen if the toilet's siphon were shaped differently, you'd learn how it relies on both atmospheric pressure and gravity. You can also learn about new gaps in your intuitive understanding when a counterfactual has a surprising effect that you hadn't anticipated. Thus, one might be surprised to learn that changing the color of cheaper patio doors from white to black is not possible because the black version of the inexpensive vinyl used in patio door construction expands too much when exposed to strong sunlight. Patio doors made of more expensive composites do not have that problem. Counterfactuals are powerful tools for uncovering or highlighting causal structure.

We frequently deploy counterfactuals in explanations to novices. The physicist George Gamow was widely known in the 1940s for his stories about a bank clerk known as Mr. Tompkins who becomes curious about physics. Gamow used the fictional Mr. Tompkins and his curiosity as a vehicle for explaining physics to the greater public. In one story, Gamow described what the world would be like if a physical constant, namely the speed of light, was changed from 186,000 miles per second to 10 miles per hour and Mr. Tomkins was present as an observer (see figure 2.2). In Gamow's words:

> A single cyclist was coming slowly down the street and, as he approached, Mr Tompkins's eyes opened wide with astonishment. For the bicycle and the young man on it were unbelievably shortened in the direction of the motion, as if seen through a cylindrical lens. The clock on the tower struck five, and the cyclist, evidently in a hurry, stepped harder on the pedals. Mr Tompkins did not notice that he gained much in speed, but, as the result of his effort, he shortened still more and went down the street looking exactly like a picture cut out of cardboard.[34]

Figure 2.2
Mr. Tompkins in a world where the speed of light is ten miles per hour. *Source*: Gamow, G. (1965). *Mr. Tompkins in Paperback*. Cambridge University Press.

Working through the consequences of this slower light speed on daily life reveals aspects of Einstein's special theory of relativity. Equally if not more dramatic experiences await Mr. Tompkins in a different world where Planck's constant, a foundational constant in quantum physics, is changed by several orders of magnitude from its true value and Mr. Tompkins gets to experience the quantum world at a macroscopic scale.

Gamow's stories describe cases of physically impossible worlds that cannot exist in any real-world scenario. Indeed, the scenarios require some glossing over of how the universe would also be quite different in many other ways. Stephen Hawking noted that the values of many physical constants cannot be changed much at all without making the universe impossible for biological organisms as we now know them to exist, a situation known as the "fine-tuned universe." This phenomenon illustrates a deeper problem with counterfactuals: It is very difficult to know all the consequences of changing one thing.

A different class of changes further illustrates the complexities of feature choices. Imagine you want to pose counterfactuals to learn more about how a grandfather clock works. You want to pick two changes that alter just one property, and you want to compare their impacts. A friend suggests that you change the color of all the pieces of the clock to bright yellow from their current shades of black and brown. A second friend suggests that you change all the metal and wood pieces to cardboard. You reject both suggestions, but for different reasons. You reject the color change because of a strong intuition that changes in the color of the parts of a mechanical system are almost always irrelevant to the workings of that system. You reject the change to cardboard because changes from hard, durable components of a mechanical system to softer, nondurable ones will almost always impair the workings of any mechanical system. Thus, both changes would be uninformative and neither would tell us anything useful about the workings of grandfather clocks in particular. Color changes will not make a difference for any device while material changes will make a difference for all devices.

Where do laypeople's intuitions about the uninformative nature of such counterfactuals come from? Most people rarely explicitly consider the utility of changing colors and materials for learning about mechanical devices, but they may all have developed more implicit senses of such patterns. In a set of studies in our lab, we showed that adults and children as young as five sensed that changes in colors were more likely to be causally central to understanding how novel animals worked than changes in size. The opposite pattern held for novel machines, for which size mattered more than color.[35] Choosing the most informative counterfactuals relies on both intuitive and explicit background knowledge of how the causal potency of different types of properties varies across different kinds of things.

The challenges of selecting the most useful counterfactuals have been recognized by scholars across the cognitive sciences ever since the philosopher David Lewis pointed out that other possible worlds were most informative when they were minimally different from reality.[36] The idea of the most informative possible world being the closest one to reality makes sense when you realize that changing too many things at once will obscure what causes what. Yet this seemingly simple strategy can quickly overwhelm human cognitive capacities. The psychologist Ruth Byrne, in reviewing the psychological roles of counterfactuals, illustrated in detail the potentially infinite sets of possible worlds that could be considered as candidates for counterfactuals in order to increase understanding of a particular entity.[37] Any attempt to consider every candidate world would quickly swamp the limits of human cognition. Humans therefore have strong biases that greatly narrow down the choices of what to change, and we tend to agree on those biases, which, in the case of events, include changing exceptional events to be more normal, making events controllable, focusing on actions, and emphasizing the most recent part of series of events that might describe a world. We don't yet fully know how adults successfully narrow down the set of possible worlds to find useful ones, but the final story may help us to more systematically explore difficult-to-explain phenomena as well as to assess the outputs of artificial intelligences.

Explanations with Mechanisms in Mind

We are now ready to delve into mechanistic explanations more deeply, especially constitutive mechanistic explanations for both artifacts and naturally occurring entities. Knowledge of such mechanisms has repeatedly been granted a special status in explanatory understanding. In particular, the centrality of constitutive mechanistic explanations in the sciences has emerged as a major theme over the past two decades in the philosophy of science.[38] Elsewhere, members of our lab and other groups have engaged in broader considerations of mechanistic explanations in human cognition.[39] Here, we'll focus on how mechanisms can foster insights.

In the philosophy of science, mechanical explanations are normally described as immersed in layered hierarchies.[40] *Mechanical* as used here refers to cases of causal interactions between solid objects, such as intermeshed gears, levers on loads, and collisions between solid objects. These kinds of mechanisms typically involve *work* in the technical sense used in physics.[41] Work, in physics, refers to the transfer of energy between objects by means of a force that follows along a path. More simply, work occurs when a force on an object causes the object to move. If you push with great vigor against a large rock and it doesn't move, no work is being done even if you are expending energy in the failed effort.

The stuff that makes up the universe—atoms, elements, compounds, solid objects, fluids, aggregates, and various other enduring matter—constitutes a far different reality from a universe that is a completely randomly distributed "pea soup" of pieces. An ever-thinning broth would be a more accurate metaphor, and, by most cosmological accounts, it may ultimately be the universe's fate as order surrenders to complete disorder and randomness.[42] But until that final ending, our interest lies in knowing how things hang together in organized ways that locally defy the progression to total disorder. Certainly, much of that organization arises from the four fundamental forces in physics: gravity, electromagnetism, the strong nuclear force, and the weak nuclear force. (There may also be a mysterious fifth force motivated by discoveries in cosmology.)[43] But beyond those basic forces, most humans see patterns relating to *causal homeostasis*, where a set of properties causally reinforce each other over time and thereby create a stable property cluster.[44] Mechanisms loom large at multiple embedded levels in such property clusters. We can construe reality as richly structured sets of layers that vary in their architecture across broad domains but that share general principles such as the relative autonomy of levels and reasonably stable manners of decomposition.

Taking a cognitive perspective, mechanistic explanations may be especially salient from infancy. A simple collision between two solid balls, as in billiards, is seen as a causal event by young infants. More than just seeing ball A causing ball B to move, in a typical collision, they see A as *launching* B in the same manner that is conveyed by the concept of doing work. We do not yet know if work-like events stand out as instances of causation at all ages and in all situations, but many such events are especially easy to notice and to employ in reasoning. Since such cases are typically observed as the movements of parts of objects, a worldwide decline in experiences with visible instances of these phenomena might matter. Of all the patterns of causation that exist in the world, those resembling instances of "work" may belong to a small set of cognitively privileged patterns that play outsized roles in later thought.

Why are mechanistic causal relations granted such a privileged position in human thought? We may be building upon a robust cognitive ability to develop representations of the spaces we encounter. Animals that move about must keep track of where they are to survive. They create cognitive maps representing their current location and their journeys. When we think about causal mechanisms, we may represent them in ways that exploit the early ability to monitor the relative positions of entities in space.[45] An understanding of mechanisms often relies on intuitive versions of work (in the technical sense used in physics). Because that sense of work is defined as the energy transferred to an object by a force along a path, it is intrinsically spatial.

We may visualize many mechanical systems in terms of objects transferring energy through their motion to other objects in a manner that serves a goal. A set of intermeshed gears does this, as does a chain of successively falling dominos. This tendency to embed work in mechanical intuitions gains support from a study in which people were shown a video of colorful novel machines with moving parts consisting of gears, levers, and pulleys.[46] People were asked to make a drawing that explained how each machine worked, or to make a drawing that would help someone else visually identify the machine. The drawings were markedly different (see figure 2.3). When asked to explain how a machine worked, the drawings often omitted, distorted, or simplified mechanistically irrelevant parts but included arrows showing the motion paths of the components. One distortion was to make the mechanical parts larger when asked to provide an explanation as opposed to a guide to visual identification. People also used more strokes of the pen in their drawings of causally relevant parts than of causally irrelevant parts. The tendency to spontaneously add arrows when trying to explain how things work has been found in other studies asking people to communicate through diagrams.[47] The modifications of visual mechanical explanations can also be construed as indicating a form of idealization or abstraction, as causally irrelevant details are omitted.

People form impressions of the mechanical complexity of objects as distinct from their visual complexity. In many cases, mechanical complexity is more influential than visual complexity. In a series of studies led by Tal Boger in our lab, we found that when people were asked to rate the complexity of familiar objects and the kind of complexity was not specified, their ratings were driven more by intuitive mechanical complexity than by intuitive visual complexity.[48] Moreover, people were better able to notice and remember objects presented visually based on mechanical complexity.

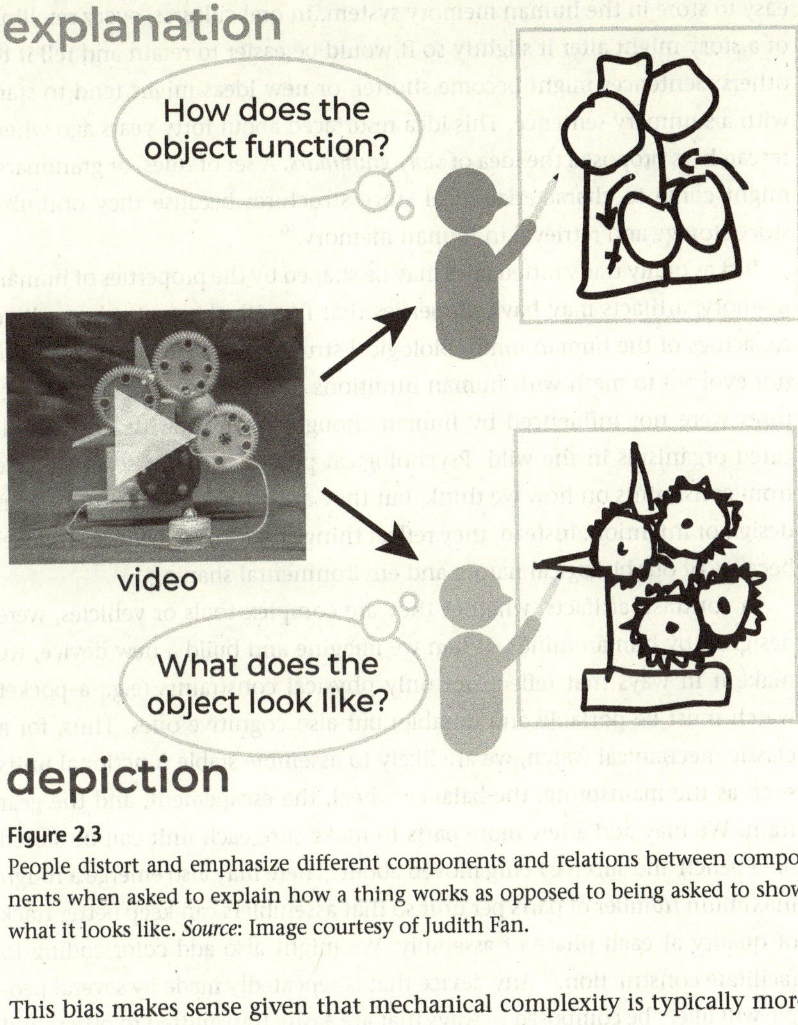

Figure 2.3
People distort and emphasize different components and relations between components when asked to explain how a thing works as opposed to being asked to show what it looks like. *Source*: Image courtesy of Judith Fan.

This bias makes sense given that mechanical complexity is typically more useful in guiding decisions on how to interact with objects.

Artifacts as Explanatory Archetypes

In 1928 a Russian scholar named Vladimir Propp published *The Morphology of the Folk Tale*.[49] Propp examined recurring structural features of folktales across several cultures. These folktales were handed down generation after generation through oral traditions, often by illiterate people. One of Propp's most intriguing claims about the structural properties of folk tales involved the gradual morphing of folktales' structure to forms that were especially easy to store in the human memory system. In oral cultures, every retelling of a story might alter it slightly so it would be easier to retain and tell it to others: Sentences might become shorter, or new ideas might tend to start with a summary sentence. This idea resurfaced about forty years ago when researchers proposed the idea of *story grammars*. A set of rules, or grammars, might come to characterize ideal story structures because they optimize story storage and retrieval in human memory.[50]

Just as orally transmitted tales may be shaped by the properties of human memory, artifacts may have properties that fit with the evolved cognitive capacities of the human mind. Biological structures and mechanisms have not evolved to mesh with human intuitions because their causal architectures were not influenced by human thought, at least with undomesticated organisms in the wild. Psychological processes might seem to arise from constraints on how we think, but they are not the product of human design or intention. Instead, they reflect things that we do, believe, and feel because of our biological nature and environmental shaping.

In contrast, artifacts, whether they are complex tools or vehicles, were designed by human minds. When we imagine and build a new device, we make it in ways that reflect not only physical constraints (e.g., a pocket watch must be portable and durable) but also cognitive ones. Thus, for a classic mechanical watch, we are likely to assemble stable functional units such as the mainspring, the balance wheel, the escapement, and the gear train. We may add a few more parts to make sure each unit can be stored on a bench and survive being moved about. There may also emerge a rough maximum number of parts per unit so that assemblers can keep better track of quality at each phase of assembly. We might also add color coding to facilitate construction.[51] Any device that is repeatedly made by several people will likely be composed in ways that are easily transmitted to others and remembered. This description of the values of stable subassemblies was eloquently described by Herbert Simon in terms of an imaginary clockmaker.[52]

When building a house, an architect may put internal bracing at equal intervals despite such intervals requiring a bit more material than unequally spaced alternatives because doing so makes it easier to keep track of components during construction and servicing. A device with many control buttons is likely to have them in spatially separated clusters of relatively few buttons so that they can be distinguished and remembered more easily than if the buttons were all in one large array. As designs are transmitted to others, they will subtly change in ways that make them easier to remember and communicate. Inspect almost any device and you will see how the contours of the mind shaped its design.

If this account is roughly correct for artifacts, children may find the interiors of classical mechanical artifacts to be especially cognitively interpretable in terms of their mechanisms. This doesn't mean that people have complete mechanistic understandings of most traditional artifacts, but they may find it easier to build such understandings for those artifacts than for other kinds of things. Artifacts may also provide compelling causal frameworks for figuring out the mechanisms underlying biological systems, nonliving natural systems such as geysers, and perhaps even psychological systems. After all, one of the most popular early accounts of how to understand the mind was Freud's explanation of mental components expressed in terms of fluid dynamics. In doing this, we may over-attribute the mechanisms of classical artifacts to natural kinds; but even with those errors, we may be able to rapidly gain causal insight into a system. We might then modify our model to make it more realistic. For example, it is common to explain how grasshoppers jump large distances in terms of their large hind legs working like a catapult.[53] But this is only a rough approximation, as catapults don't throw themselves through the air. By forcing organisms and systems into available and accessible mechanistic models, we may quickly grasp rough causal models, but for most natural kinds, it will be an imperfect alignment.

As AIs occupy larger and larger roles in artifact design, those designs may well become less comprehensible when they no longer are shaped by what human minds find easy to represent. It may not be feasible to force current AIs to make designs that conform to more naturally human ways of thinking without severely hobbling their performance. If so, we will run an increasing risk of being unable to anticipate potential consequences of those designs.

Mechanisms in Biology, Psychology, and Physics

In the mature sciences and engineering, we find consistent differences across research areas in terms of typical causal structures of interest. The best-fitting explanations vary across fields as they reflect real-world differences in causal patterns. At the highest level, virtually all artifacts and organisms share the same principle of being local instances of reverse entropy. *Local* is an essential qualifier considering the second law of thermodynamics, according to which, globally, entropy must always increase. The universe is always proceeding toward greater disorder and randomness. Entities like clocks and amoebas are local cases where random collections of molecules temporarily become more ordered, but only by creating greater disorder elsewhere (e.g., through the combustion of fossil fuels or the metabolization of glucose).

Beyond this overarching commonality across virtually all entities, large domains such as biology, physics, and psychology have their own distinctive causal signatures. In physical mechanics, contact between two objects and instant consequences are usually necessary parts of coherent events. Billiard balls don't act on other billiard balls at a distance, and the motion of one ball hit by another ball doesn't start after a three-second delay of no visible changes. But at the psychological level, most causal interactions occur without direct physical contact and often involve substantial temporal delays.

Explanations about organisms differ in several ways from those about devices. For example, it is rare to ask about the function of an entire organism, especially if it is not domesticated. It is commonplace, however, to ask about the functions of parts of organisms, such as skeletons and organs. In contrast, the function of an entire device is often the first part of an explanation about it. Thus you might ask about the function of humps on camels, but far less often about the functions of whole camels. In contrast, you would likely ask about the function of a tool before asking about the functions of its parts.[54]

Consider in more detail the way explanations vary across the domains of artifacts and biology. The hierarchy of function, subfunction, and mechanism that is so compelling for many complex devices is less clear-cut in both biology and psychology. A refrigerator may be neatly decomposed into a few high-level components such as the compressor, the condenser coils, the cooling coils, and the expansion device. These then are further decomposed neatly into subunits, such as those making up a compressor. Similarly, an internal combustion engine decomposes into the drive shaft, fuel injectors, piston chambers, spark plugs, and other such units. Each of these then further decomposes into functional and mechanistic combinations.

The boundary between mechanism and function can become blurred when functions become sufficiently embedded in a system. A refrigerator compressor can be defined in functional terms—what it does—as any device that compresses the refrigerant regardless of how it works in detail. It can be considered in mechanistic terms—how it works—when describing how its outputs and inputs are causally involved in the refrigerator itself or when the compressor is described in terms of the causal interactions of its parts. When an entity is multiply realizable in different physical instantiations, it can also usually be described in functional terms.[55]

In biology, it is tempting to assume the same clean hierarchical structure we find in most artifacts, where the interplay between function and mechanism can be neatly disassembled. But organisms are rarely that straightforward. More than sixty years ago, in an influential essay comparing physics to biology, the prominent evolutionary biologist Ernst Mayr denounced "Descartes's grossly mechanistic interpretation of life."[56] Biological systems often include vestigial or local "good enough" subunits that persevere. In successive artifact designs, in contrast, such units are usually discarded and replaced by more efficient parts.[57] The human appendix is a well-known example. Evolution by natural selection is not able to backtrack and undo inefficient or unnecessary components.[58] Those components may gradually wither away, but often leave traces for many generations. They may also be repurposed in a nonoptimal but sufficient manner. Biological causal structures also tend to have more cycles and other processes that depart from the simple ideas of mechanical work. In other areas of biology, such as the human immune system, the overall organization defies any kind of elegant hierarchical unpacking of functional and mechanistic units. Notions such as hierarchical control are replaced by networks of interactions. The biologist John Trowsdale argues that, while analogies to physical mechanical systems may work in explanations of systems like the heart, lungs, and kidneys, the best analogies for the immune system tend to use examples from warfare or other social interactions.[59]

Scholars continue to debate what should count as instances in the full range of mechanistic explanations.[60] Do cycles and feedback loops always qualify? How probabilistic can causal links be and still be considered mechanistic? These and many other questions are not central to our analysis of insight, however, so we'll focus primarily on deterministic interactions between solid component parts that are performing work.

In short, explanations have rich and multifaceted structures, and those structures influence how satisfying certain types of explanations are for a particular domain. Context can strongly influence those preferences; context, in turn, is heavily influenced by background knowledge. If children today are relying on different forms of background knowledge from children fifty years ago, such as encountering far fewer revealing interiors, we might also expect differences in how they seek out and choose among explanations.

Mechanisms, Biological Thought, and Research Communities

In the 1950 and 1960s, most universities had either a single biology department or two distinct departments of botany and zoology. Today, many universities have divisions of biological sciences comprising several smaller departments. Other universities have decided to organize two departments: one focusing on integrative biology, or evolutionary biology and ecology, and another focusing on molecular biology and, often, development. Botany and zoology have been combined again. These fissions and fusions of areas within natural science departments are atypical when compared with other sciences.

Why is biology different? The most relevant factor concerns the typical explanations provided in different areas of biology. Areas commonly described as molecular biology or cellular and developmental biology gravitate toward mechanistic explanations. In contrast, departments often known today as evolutionary biology and ecology, or integrative biology, instead employ explanations resembling unique historical descriptions or complex integrated systems that defy easy decomposition into parts. For example, a researcher in the molecular biology area might focus on the cellular components involved in mitosis and how those parts caused the chromatids to line up so precisely during the metaphase.[61] Another might study how individual cells are capable of movement. Their studies explore how certain signals cause the buildup of specific molecular structures on one side of the cell that result in the cell wall moving outward while other signals cause the cell to reabsorb comparable structures on the trailing edge for reuse and deployment at future leading edges.[62]

In contrast, an evolutionary biologist might argue, based on fossil records, that epigenetic events were supported by historical periods of certain kinds of change in the evolution of plants.[63] Another might study how the arrival of a new environmental threat to a species, such as a new predator of nesting birds in the arctic, resulted in the emergence of several different subspecies adapted to deal with the threat. The research could not predict exactly what adaptations would emerge, but might be able to show post hoc how certain conditions strongly pushed a species toward a class of different structures that all serve the common function of protecting nests.

I was blissfully unaware of this contrast in explanatory styles and the tension it could create when I was a biology major at MIT from 1969 to 1973. In retrospect, my ignorance was not surprising. In the many biology courses that I took, the largest organism we ever discussed was *E. coli* (*Escherichia coli*), a family of bacteria with many strains that live symbiotically in mammalian guts, but which is best known for variants that can contaminate food and be poisonous. In my third year, at the last session of a lab course on molecular mechanisms, the department chair dropped by to tell us that some people might think it was rather odd that as biology majors we had not studied any organisms larger than *E. coli*, "even though the fundamental mechanisms are all the same." He had therefore decided we should encounter something bigger. A cart was rolled into the lab piled high with fetal pigs, one for each of us. We were told to "dissect away," and nothing else. As the teaching assistants (who all were molecular biologists) wandered helplessly about, watching us, we all started picking apart our pigs as an ungraded exercise. It seemed utterly disconnected from everything else I had learned in the department. In my courses, the focus on mechanism was clear from the start. Introductory biology was a magnificent series of lectures on molecular mechanisms. Everything seemed to work like marvelous clockwork. We never even thought about evolution or ecology.

In the following decades, even a novice could see that the two areas of biology were overlapping and intersecting to ever increasing extents. Fascinating lines of work emerged on the commonalities and differences in viral and bacterial evolution, and classic evolutionary accounts started to study relations between genetic sequences and other work done with fossils.

Biologists and philosophers of science are now evaluating the limitations of both explanatory styles for both domains. Does a mechanistic stance distort or even miss certain patterns? Most discussions warn about equating biological systems too closely with classical machines.[64] One review, while conceding the heuristic value of such comparisons, also contained several cautions that are widely echoed in other analyses. For example, the authors stated, "Notoriously, evolution is an opportunistic tinkerer, blindly stumbling on 'designs' that no sensible engineer would come up with."[65]

Are there more intricate senses of "mechanism" that can solve the problem of imperfect fits with biology? It is tempting to broaden the meanings of *mechanism* to include almost any stable causal pattern with a discernible structure, but including so many senses often results in scientists talking past each other because they have different notions of mechanism in mind. This has led to prominent calls for more care and rigor in the use of the term *mechanism* in the life sciences.[66] Here, using it in the sense of an ordered progression of parts doing work, we explore how humans may be both gifted and hindered by it. We may be gifted when analogies to machines allow us to use a powerful, spatially enhanced way of noting and remembering structured causal clusters. But we may be hindered when making unwarranted inferences about causal patterns that don't exist, such as assuming unrealistic levels of determinism or exaggerating the decomposability of a system into stable subassemblies with identifiable functions.[67]

The value of applying mechanistic notions to biology seems to depend on the kind of biological process involved and its level of granularity. Mechanistic descriptions may work best at the microscopic level and worst at the level of a whole organism. Indeed, one of the most compelling and insightful cases of thinking of a biological entity as a machine is in terms of the flagellar motors found in many microorganisms. Parts of mechanical motors map in sensible ways onto biological versions that are described as rotors, stators, hooks, universal joints, neck linkers, and switches.[68] All such decompositions continue to be debated by stressing the enormous disparities of scale and how scale changes result in different influences of environmental factors, such as temperature. Yet several reviews by leading scientists use mechanical metaphors extensively and in ways suggesting consensus on what they mean. In the end, a fascinating project lies ahead at the intersection of biology, philosophy, and cognitive science that addresses the following question: How should we carve up the many areas of biology to capture those aspects that effectively enable human minds to reason about causal patterns while avoiding aspects where such metaphors do more harm than good?

* * *

Humans have a unique hunger for insightful explanations. This drive to make sense of the world is especially strong when it seeks answers to how and why questions. Explanations come in several types and vary in their properties in ways that resonate differentially with broad categories in the world such as biological kinds, nonliving natural kinds, and artifacts.

Adults show impressive sensitivity to these patterns by preferring certain explanatory structures and kinds of content in ways that vary systematically across broad domains. Explanations also distort actual causal patterns in the world because of imperfect fits with highly available explanatory schemes and because of necessary abstractions and idealizations.

Mechanistic explanations have figured prominently in our discussion of the ways we use and understand explanations. Our inclination toward such explanations may arise from how they can build on the ways in which we automatically represent objects and their relations in spatial layouts. Artifacts, construed as things made with purposes and composed of purposeful parts, may foster the growth of early mechanistic understandings. Traditionally they have been adapted to fit with the natural contours of human minds and may therefore have provided especially powerful insights to young humans encountering them. But a dark cloud has been hovering over this happy harmony between the things we make and how we think: the possibility that artifacts have broken this harmony because their working parts have become causally opaque. Chapter 3 examines this claim, considers the supporting evidence, and characterizes the kinds of changes involved. It thereby creates a foundation for the examination in chapter 4 of how pervasive changes in artifacts might alter our preferences for understanding and explaining reality.

II Purposeful Parts

3 Disappearing Details

In 1968, Sharon Sites Adams set out alone on a twenty-five-foot boat to sail from Los Angeles to Hawaii. She was the first woman to complete the trip alone. No one should underestimate the challenges of hitting that target. Reaching Honolulu from Los Angeles requires maintaining a precise compass bearing of 249.97°. Straying even a few degrees from that bearing is to court disaster. Given wind, currents, and waves, it is impossible to stay on course without navigational tools. The navigational challenges were painfully illustrated by the disastrous Dole Derby in 1927, the first airplane race to Hawaii.[1] Of an initial eighteen entrants, only two pilots were successful. Some planes and crews simply disappeared without a trace in the vast Pacific. The pilots often couldn't see the horizon in sextant sightings because of cloud cover and instead had to rely on less accurate bubble levels similar to those used by carpenters.

Adams used the same tools that had been used by sailors for more than two hundred years: a gimbaled compass, a taffrail log, a chronometer, and a sextant. After a few months of tutoring, she learned how to find her location on charts. The taffrail log, which started appearing on boats in the late eighteenth century, is the most mechanistically intuitive (see figure 3.1). They measure distance traveled and come in sturdy wooden boxes with enticing components, such as a fluted brass cone, called the spinner, with propeller-like vanes. The spinner is tied to ten meters of tightly woven line attached to a round meter mounted with a special fitting on a wood railing at the stern of the boat (the taffrail). The line must be tightly woven to maintain a consistent resistance when twisted. The meter has a needle showing the distance traveled in nautical miles. (The log is surprisingly accurate and reliable, except for those occasions when a shark bites off the spinner.) Information from the log can be integrated with information from a magnetic compass, chronometer, and sextant. Each has to be understood in terms of how it works. The compass, for example, must be "swung" periodically to measure how the ferrous metals and wiring on the boat might be distorting its readings. The sextant is used to estimate the angle between the horizon and the stars, the sun and the moon. Finally, the chronometer is an especially reliable and accurate clock that enables the navigator to figure out their longitude from the other readings.

In 2023, the same trip from Los Angeles to Honolulu can be guided by a compact handheld device: a GPS unit with integrated marine charts. The unit shows where the boat is on a chart, its heading, and its speed. The user doesn't need to know anything about how the unit works. It is much more accurate than the tools used by Adams, especially in dark and stormy conditions. It is also much cheaper and requires minimal expertise. Only the sentimental or foolhardy would choose to navigate a long ocean trip the old way. Yet the experience is completely different. In the older version, you were acutely aware in real time of the outputs of all your devices. You knew how they worked and how those mechanisms could lead to specific distortions. You were actively interpreting and integrating all outputs. You

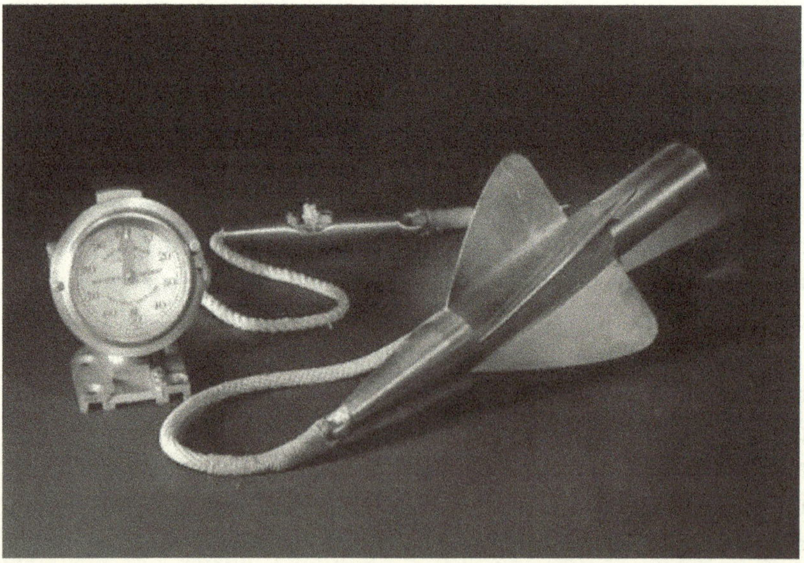

Figure 3.1
A Walker taffrail log, first made in 1877. *Source*: Image courtesy of the Smithsonian Institute.

felt immersed in the whole system as an engaged participant. Today, you just look at a small box in your hand with a digital chart displaying an arrow corresponding to your boat. Some people just use their cell phones with a marine navigation app.

Over the past few decades, a profound change has occurred in how to navigate a small boat on offshore passages. Navigators in 1960 had well-grounded understandings of how several devices worked and could collectively be used to determine location. They had insight into what it meant to navigate. Their 2023 counterparts simply need to view a screen and follow the arrow. Those who mastered earlier ways of navigation and still have the instruments might well be invaluable when a boat's GPS fails or when GPS signals are jammed by regional conflicts, but few mariners today ever consider such risks.

GPS systems do occasionally fail and make trips more challenging. About twenty-five years ago my son Dylan and I took a small sailboat for an eighteen-hour wet and windy passage along the New England coast from Rhode Island to our home in Connecticut. As we approached our home harbor in darkness and fog, we discovered that our handheld GPS had failed. (Phones at that time had no GPS mapping functions.) We slowed to a snail's pace, looking and listening for the many rocks and shoals in the area; then an enormous bonfire appeared off to our right. It just happened to be one of our neighbors who was belatedly celebrating the Fourth of July on his beach. That fire saved us hours of tense creeping in darkness and was a vivid reminder of the value of traditional means of navigation.

The immersion of navigators into a web of instruments does not mean they fully understand all the relevant mechanisms of the stars, the earth, and the oceans. Sharon Sites Adams became a skilled user of her navigation instruments, and she understood how each one worked in a proximal sense, but her knowledge of astronomical distances was surely not complete. A course of celestial navigation for mariners is highly focused on the task at hand: getting to a new location on earth over water accurately. It does not care much about interstellar distances. Thus, in the 1960s, celestial navigators assumed a "heavenly sphere" model in which stars, planets, the sun, and the moon were located on a sphere at an infinite distance from earth. By treating the positions of celestial bodies as fixed points on an imaginary sphere, various numerical manipulations became much easier. Other calculations were simplified by assuming the navigator was an observer who was standing at the earth's center.

In this respect, Western navigators were not so different from the legendary Polynesian navigators who achieved remarkable successes over vast stretches of the Pacific.[2] They also used fictional models, such as moving islands and invariant star lines in the sky. The similarities were even greater prior to the mid-eighteenth century, when accurate chronometers first enabled people to calculate longitude. Before those chronometers, everyone relied on variations of dead reckoning plus rules of thumb such as sailing along a line of latitude where certain destinations were known to be located. This practice of "running down the latitude" worked because it was relatively easy to measure latitude with instruments such as sextants or cross-staffs. These earlier voyagers knew much about the causal roles of their instruments, their maps, and other navigational aids.

We normally operate with incomplete understandings and often don't realize how little we know. Yet, with all these limitations, we can retain strong and legitimate senses of cognitive agency and a deep appreciation of how we are embedded, active participants in an elaborate network of causal patterns. One of the great human skills is thriving despite such limitations. We can all appreciate the valuable knowledge of navigators in comparison to astrologers.

One could take Sites Adams's trip as a quaint vignette about a niche skill that is no longer relevant, but the underlying message is broader. Nautical navigation is just one of thousands of examples of a profound recent revolution in how we experience and use technologies. If mechanisms are disappearing into mystery, are there cognitive consequences for children growing up in those settings? The answer depends on whether children normally obtain cognitive benefits from revealed mechanisms or whether that information goes over their heads. While traditional views of children's minds saw them as oblivious to underlying causal relations, in several recent studies, even preschoolers track and remember much about these patterns. Do their abilities to learn such patterns have lasting cognitive benefits? To better address these questions, we must first characterize historical changes in artifacts in more detail.

From Transparency to Opacity

Most everyday artifacts of any complexity have changed dramatically over the past few decades. Cars, clocks, cameras, phones, light bulbs, and countless other devices may function in roughly the same ways, but their enabling mechanisms are radically different. Modern technology makes devices more versatile, reliable, and precise than their predecessors, but often at a cost of becoming less explainable.

Concrete examples of disappearing mechanisms are everywhere. For example, table 3.1 lists three common devices whose workings have changed from the intuitive to the mysterious while providing the same function and often retaining the same surface appearance. These are just a few examples of the many artifacts whose primary working parts have transitioned from readily apparent mechanics to opaque circuit boards. In other cases, simple electronic circuits of a dozen or fewer components have turned into complex circuit boards. For example, incandescent light bulbs, which function based on a current running through a wire filament in a vacuum, have ceded ground to the circuits underlying LED light bulbs even as they may be similar on the surface. With an incandescent bulb, you could visualize current as a fluid flowing through the narrow filament. When the flow was restricted, the wire heated up from "friction" and emitted light. With LED bulbs, a current runs through a circuit consisting of several diodes in which the dynamics that produce light are difficult to describe and visualize.[3] How are today's children to gain any sense of how things work? What analogies and metaphors to mechanical models can they draw upon?

I recently learned how circuits and chips stabilize a cell phone camera by using vibrating plates that employ stabilizing effects like those in gyroscopes. Without that prior model of gyroscopes, the plates would be confusing. When the world offers fewer and fewer glimpses of easy to access mechanisms, children growing up in that world may have impoverished cognitive "tool kits" for deciphering what is happening around them.

Today most things cannot be opened to see how they work. A few decades ago, once you removed the relevant covers, you could grasp how cars changed speeds, toasters adjusted browning, and alarm clocks rang. Many things could be repaired after a simple inspection of their insides or, at worst, by consulting repair manuals with line drawings. To see the change in more concrete detail, I start with an especially mechanistically transparent device from my childhood: an outboard motor.

In 1960, my father bought a 2 hp British Seagull (right side of figure 3.2), our first and only outboard motor. It was about as mechanically transparent as a gas-powered outboard could get, and for good reason, as it broke down

Table 3.1
Similar on the Surface; Radically Different Underneath

	1950	2024
Wall clock	Spring, balance wheels, escapement, key winding component with ratchet, many gears	Circuit board and digital display mimicking a clock face, batteries, wiring
Cash register	Springs, levers, gears, paper roll, ink ribbon, numbered keys, rotating drums, bell	Circuit board and digital display, touch screen, wiring, barcode scanner, contactless pay option
Slot machine	Wheels with icons, springs, levers, axles, and many other moving metal pieces; clearly worked as a stand-alone unit	Circuit board and digital display creating the illusion of a locally controlled mechanical wheel that spins; internet connection to a central server that sends images masquerading as locally triggered spins

Although modern versions of these three devices can come in forms whose surface appearances are similar to their 1950s counterparts, their innards are both remarkably different and surprisingly like each other across various modern devices.

continuously. (Repair problems may have led to a subtle change in its marketing slogan from "The best outboard motor in the world" to "The best outboard motor *for* the world" (italics added). But Seagulls were easy to take apart and repair, even on the water in a small rocking dinghy. Seagull owners were distinctly proud of their abilities to keep their outboards running year after year. The "Saving old Seagulls" online group remains active and lively more than thirty years after the company ceased production.[4] (Our family finally gave our Seagull away after it broke down three times on the same weekend.) In contrast, a popular modern counterpart is the Torqeedo 2hp electric outboard. Its innards are opaque and the only thing an owner can do is replace the battery (left side of figure 3.2). No online repair manuals or approved diagrams of parts are available. The only relevant material is an operator's manual. Torqeedos are much more reliable—and quieter—than Seagulls, but they are mechanistically uninformative.

The mid-twentieth-century world wore its mechanisms on its sleeves, or at least right under them. Now most workings are hidden in circuits. These buried mechanisms disable learning. It isn't possible to look more closely at how parts interact or to tinker with specific parts and subassemblies. Recall,

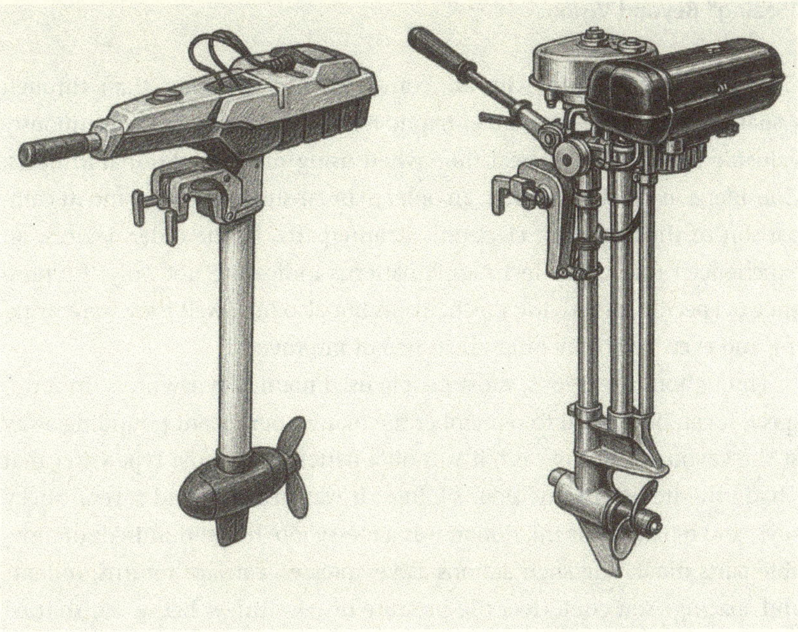

Figure 3.2
A small boat outboard motor from today (*left*) and from 1960 (*right*). The British Seagull on the right could be easily taken apart into every moving piece and repaired. Even for outboards of its day, the Seagull was unusually mechanistically transparent. The Torqeedo on the left can only have its battery removed and replaced. *Source*: author.

for a moment, the pleasure of insight upon first looking inside a classic music box. You immediately see how the rotating drum with oddly spaced bumps plucks separate metal fingers of a steel comb to make distinct tones. No similar rush of insight accompanies ripping apart a greeting card in a fruitless attempt to understand how it plays a song. When you finally see what is hidden in the card you find innards that look the same as those of hundreds of other devices performing completely different functions, such as smoke alarms, stud detectors, garage door openers, and television remote controls. You often can figure out the battery and perhaps an input sensor and an output part, such as a speaker. The remainder is usually one or more integrated circuit chips whose functions are unknown. How is such a device causally related to the production of music? Beyond the idea that the battery provides the power, most people would have nothing else to say.

"Seeing" Beyond Vision

Cues to internal complexity are conveyed in ways other than through visual inspection. For example, haptic feedback can provide continuously valuable information in real time when using many traditional artifacts. Consider a manual typewriter, an older manual-shift car, or a piano in comparison to their modern electronic counterparts. In the older devices, an experienced user could feel haptic patterns indicating not only the presence of specific underlying mechanisms but also how well they were working and even how they might be tuned or improved.

Throughout the 1960s, most people used manual typewriters. In retrospect, it can be painful to remember the many hours spent pounding away at the keyboard. In my case, it was on a battered Olympia typewriter that I had inherited from my older siblings. It was heavy, it had several sticky keys, and replacing its ink ribbon was a messy job. It had hundreds of movable parts mediating such actions as key presses, carriage returns, indents and spacing. You could feel the pressure of your finger being transmitted through the linkage to the printed page. You could tell if you hadn't typed strongly enough without looking. You could tell if some element in the causal chain from key to page strike was impaired and often what part it was. You could usually tell from touch and sound what kind of paper you were typing on (e.g., bond, mimeo, or carbon copy). A small number of writers today swear by their manual typewriters, but most people who lived through those times have no interest in abandoning their current keyboard and computer and returning to the manual machine. Why, then, does it matter if later generations never learned the touch and feel of a clearly obsolete mechanism that they are very unlikely to ever use?

Are there enduring cognitive benefits from having haptic senses of how manual typewriters worked? If so, they must be ones unrelated to using those typewriters again in the future. One possible advantage might be having a more refined sense of how physical actions can be amplified in one way by trading off for something else. Thus, you might have a sense of how tapping more firmly on a key caused the typebar to move faster, which in turn caused a darker imprint on the page. Greater pressure is transformed into greater speed, which is transformed back into greater pressure on the ink ribbon and then onto the page. You see how this is achieved through a connected chain of metal levers and pivot points. Having a refined sense of this causal path might help people better understand older printing presses as well as more contemporary devices such as bicycle derailleurs and car jacks. Users may also have a better sense of typefaces and how they have evolved, and why most typewriters have the peculiar "qwerty" sequence.[5] They also learned how different paper media reacted to different kinds of inks and pressures, which may enhance insights into making signs and posters. An experienced manual typist might learn to associate certain sounds with kinds of media that you might be typing on. More expert typists who worked in typing pools could recognize which person was currently typing by just listening to the cadence of keystrokes.[6]

Users of manual typewriters may have also learned more general and enduringly useful lessons. They gained a sense of the complexity of manual typewriters in comparison to many other devices. They could see how form and function interrelated and see more easily what parts should be routinely serviced and why. They easily distinguished problems arising from wear and tear from those occurring through accumulation of dirt and detritus. Such insights could easily extend far beyond typewriters.

What does a person today know about how a keyboard is connected to words on a sheet of paper? If they know anything mechanical, it is probably because of an unusual problem such as the faulty butterfly keyboard mechanism in many Apple laptops made from 2015 to 2019.[7] They probably know nothing about the basis of that failure. Beyond that, people have no intuitions about how their finger movements on a keyboard are related to a printed product beyond knowing that somewhere there must be a computer to store the information and a printer to put it into hard-copy form.

Some sensations of manual typewriter keyboards are present in current computer keyboards. They provide feedback indicating, often with a click, that a keypress has been registered. Manual typewriters did this by virtue of how they worked mechanically. Electronic keyboards do not give tactile feedback to mimic manual typewriters that most current users have never used. Instead, people can use a keyboard more efficiently with feedback telling them their keystroke has been registered.[8]

Consider now a 1960 Volkswagen Beetle. It had a stick shift and no power assists for either the brakes or the steering. A driver could feel the road—not just the big bumps but even subtle changes in gravel density. If the steering felt loose and sloppy, you knew it probably was caused by a problem in one of the tie rods (either a loose nut or damage to the rod itself). If manual shifting was becoming difficult, you could usually tell if there was too much slack in the clutch pedal or a loose bushing. Different vibrations and sounds from the air-cooled engine could suggest compression problems in a cylinder or spark plug timing issues. I can think of these potential problems more than fifty years after last driving such a car, which certainly suggests that they were cognitively robust and well preserved. Haptic, auditory, and even olfactory feedback constantly reminded an experienced driver of the moving parts surrounding them and provided a sense of how well they were working.

Finally we turn to pianos, such as a concert grand. Even the most experienced pianists may not know the specifics of each mechanical component. A grand piano can have over twelve thousand individual parts. Each key has thirty-six adjustable parts linking the initial key press to the sound that emerges. Master pianists who don't know these details do still have a refined sense of many components and of how haptic information implies different kinds of problems. Concert pianists can tell instantly when something is wrong and can describe the problem in ways that enable a piano technician to identify what needs to be adjusted. If they say that the action is too stiff, expert piano technicians know that, for that pianist, it means that certain parts need to be lubricated. For another pianist, it might mean that the depth to which the key goes down needs to be adjusted. The pianists may not know what each part does and what it looks like, but they know the sensations corresponding to different parts that need to be adjusted by master technicians.

Other concert pianists can tell precisely what component should be adjusted and why. Mitsuko Uchida is known for having a detailed understanding of the mechanics of pianos. She prefers to take her own piano with her when she travels. Her grasp of the mechanics comes across clearly in this excerpt from an interview: "You get stuck when the weight is different key to key, the piano has been sloppily prepared, and the dampers have not been adjusted—or the spring in the pedal. . . . Problems can emerge when the pin underneath the key is dirty, or the other pin in the middle of the mechanism is dirty, or rubbing."[9] In other studies, haptic information helps experts identify the lowest compressibility of the parts of action, free-travel times of the hammer, and the maximum hammer velocity trajectory.[10]

Almost any setting in which a person uses a mechanical device to achieve a goal provides multisensory cues to the mechanical components and how well they are working. More expert users of those devices can be remarkably sophisticated monitors of the status of each of hundreds of parts. In addition to typewriters, cars, and pianos, the same sense of mechanical parts and their status occurs with small sailboats, manual looms, and most mechanical devices that a user operates directly. Analogous situations may be present when humans use living things as "tools," such as when riding a horse for transportation. Experienced riders can sense problems in biomechanical components in a horse's body, such as injured joints, muscles, and tendons.[11]

These examples may seem related to theories such as *embodied cognition* and *embedded cognition,* but they are importantly different. Embodied cognition generally refers to the idea that an organism's whole body is intrinsically part of higher-level thought. Thus, your physical body, your movements, and how they are involved in a specific situation are thought to be intimately related to how your concepts are represented and employed in reasoning.[12] Embedded cognition focuses on how the body interacts in real time with the environment, usually in the form of artifacts or other agents. In both cases, individuals seem to be merged with the environment, making the boundaries between them vague and indeterminate.

Here, however, we focus on a heightened appreciation of the mechanical workings of artifacts and even of some living things, such as horses under saddle. Agents are fully aware of the objects they are acting upon and are not subordinating themselves to them. Instead, they are achieving a heightened awareness of the causal processes intrinsic to the artifact or organism by interacting with it as an agent rather than merely observing it or being immersed in it. The sense can be thought of as *instrumental cognition.* Instrumental cognition is an enhanced appreciation of the working details of entities through their active use. It usually involves multiple perceptual systems. Devices become more vivid rather than disappearing in a blurry blend with the mind. Most musical instruments are good examples; personal vehicles once were, as well.

The benefits of instrumental cognition arise only when there are ample clues to internal causal patterns, usually of a mechanical nature. Electronic pianos and fly-by-wire jets tend to thwart such insights. Fly-by-wire jets can pose serious risks for pilots accustomed to mechanical systems, whose intuitions are ill-suited for electronic ones.[13] Several makers of such electronic devices, such as pianos, have tried to add back haptic sensations to help users get some feedback on their physical actions.[14]

Patterns of Parts Changes

The primary factor behind changes in the parts of more complex multipart devices has been to replace mechanical moving parts with solid-state integrated circuits wherever feasible. In most cases these changes soon resulted in versions that were less expensive and more reliable and that required less maintenance. Over the extraordinary diversity of complex artifacts throughout the world, the shift from mechanical innards to electronic ones has resulted in similar changes in their parts.

Parts got smaller. Many parts of an integrated circuit board are so small that they cannot be seen without assistance. The smallest directly visible elements of any entity are about the width of a fine human hair. As of 2023, distinct functional units in a cell phone are many thousands of times smaller than human hairs, as little as five nanometers across—only ten times larger than some atoms. These are not even visible with an optical microscope. They are hardly traditional physical objects at all, since they are visually undetectable and can be influenced by quantum effects such as tunnelling (when objects seem to pass through each other), which is far different from classical Newtonian mechanics.

Parts can't be found. Parts obviously cannot be seen if they are too small. Even larger parts, when embedded in blocks of plastic, can be impossible to locate, and the remaining physical mechanical parts of mostly electronic devices may not be isolatable. The tiny gyroscopes used in cell phones to stabilize images easily evade detection.

Parts are not separately removable. Modern electronic devices often are connected to each other in ways that make it impossible to isolate them. They may be embedded in a circuit consisting of chips with conducting channels layered on a thin wafer of silicon. Chips may then be irreversibly embedded in circuit boards. Such parts cannot be replaced or temporarily isolated to learn more about their functional roles.

Parts are not manipulatable. Parts cannot be physically manipulated in situ, thereby defying attempts to tinker with a part to explore its properties. Often the only manipulation one can do is to irreparably destroy the larger entity in which the parts are embedded.

Parts don't have affordances. Traditional mechanical parts often immediately reveal how they work. You can see levers that provide mechanical advantage, springs that store pushing energy, and so on. Those insights help us figure out how parts work together. Today's circuit boards look the same across diverse devices; you may have no idea of what the device is once its exterior shell is removed. The property of objects revealing their usefulness to humans and other animals was called an *affordance* by the psychologist J. J. Gibson.[15] It could be as simple as a smooth, hard surface affording a better walking path and as complex as piece of furniture affording adjustable reclining. One could directly perceive how an entity provided certain opportunities. (Typically, affordances refer to relations between animals and objects; here a broader sense is suggested so as to include relations between parts within a designed artifact, such as how a lever enables a spring to move a gear.)

Insides are more similar in appearance across drastically different devices. Consider the insides of a digital clock, a smoke alarm, a wireless router, and a handheld GPS device. They do entirely different things and can look quite different on the surface. But put their internal parts on a table and they may all look the same: perhaps half a dozen black rectangular chips stuck to a green plastic board with silver-looking channels connecting them, along with a smattering of smaller objects on the green board. Without any exterior cues, how can they be distinguished, let alone be observed in real time to learn how they work?

Parts oscillate faster. The fundamental frequency, or basic speed of vibration, increases as objects shrink in size. The cognitive scientist Herbert Simon highlighted this pattern in discussions of how complex systems become stably organized.[16] Its relevance to visually apparent mechanisms can be seen with the use of resonant frequencies to keep time in watches. A classic pocket watch used a balance wheel with an oscillation rate of two cycles per second. If you opened such a watch, the role of the balance wheel was obvious. It was even easier with the pendulums of a grandfather clock, which typically operated at one cycle per second. In contrast, modern purely mechanical timepieces (usually serving more as markers of affluence) can have equivalents of balance wheels that oscillate at more than twenty thousand times per second. At such speeds, the motion and its consequences cannot be seen with the naked eye. Inexpensive electronic watches and clocks vibrate at around forty thousand beats per second, and it is impossible to observe how those vibrations are mechanically linked to the hands of a watch. Smaller is faster. A leading technology developer, Mustafa Suleyman, recently described a future with an "atomic-scale nanomotor" that could rotate forty-eight billion times a minute.[17]

Shortened lifespans. A subtle but often consequential change in many modern devices arose from the increasing difficulties in identifying and removing parts: The ability to repair many devices plummeted, or repairing a device became much more expensive than simply buying a new one. With new twenty-four-inch televisions costing as little as sixty-five dollars, even the most sentimental owner is unlikely to take their five-year-old broken television to a repair shop (if there still is one anywhere nearby). As consumers shifted to replacement rather than repair, manufacturers naturally made component parts that were not designed to last beyond a typical replacement cycle. With a few notable exceptions where a product is fully covered by a "lifetime" warranty, parts are no longer designed to remain useful over several decades. This trend fits well with the "planned obsolescence" of many devices, which many scholars see showing a dramatic rise since the late 1950s.[18] Planned obsolescence implies a deliberate program to make devices obsolete, but it also may have been facilitated less intentionally as the nature of parts changed.

Mysterious causal interactions. Classic devices reveal linear chains of events, often in the form of doing "work." As described earlier, *work* means the transfer of energy from one object to another by a force moving through space. We think of this in terms of such actions as pushing, pulling, launching, and triggering. Work is difficult to discern in the crisscrossing mazes of electronic circuits. At small enough scales in circuits, tracking the "causal flow" seems almost meaningless.

In short, children today across the economic spectrum and throughout much of the world have radically different developmental experiences with underlying parts than their parents and grandparents did. Tinkering with parts has largely become impossible. We cannot observe why things fail or how they can be fixed. With few exceptions, those experiences today are more impoverished than they were a few decades earlier. If a persistent child finally finds a way to take apart a modern device and peer inside, its interior provides no insights. Parts often can't be identified, tweaked, or easily removed. Parts can be embedded in blocks of plastic or sealed containers. Modern electronics make it impossible to isolate many meaningful units.

Taken together, these differences could feasibly affect children's and adults' understandings of the world around them. The impact might be substantial for everyone, or it may be confined to just those individuals who are especially intrigued by mechanistic explanations. Some people can't help but pull things apart and tinker with them, while others seem more interested in sharing stories about other people, and still others are obsessed with sports. To what extent are the changing components of devices having an influence on children and adults who seem much less interested in how things work? I will suggest that, as communities, we all benefit from having a shared understanding of the workings of the world and feeling the thrill of gaining new insights and the accompanying sense of empowerment. Chapter 5 and later chapters provide a better sense of just how detailed a widely shared understanding might be.

Adding to the confusion, many modern toys have gratuitous solid-state chips responsible for properties irrelevant to their primary functions: toy hammers that flash when swung, or foam-ball shooters that produce gun sounds. How are today's children to gain any sense of *how*? When the world not only offers fewer and fewer glimpses of easy-to-access mechanisms but also provides misleading distractors, children growing up in that world might come to have especially stunted cognitive tool kits for deciphering what is happening all around them.

Electrical Devices

The change in purely mechanical devices was recapitulated with the emergence of the first electrical devices and the dramatic transformations they soon underwent. This may all seem like long-ago history at first, but it is surprisingly recent and rapid. Moreover, the earliest electrical appliances were usually mechanically intuitive.

My father was born in 1910. As a young child, he never encountered electrical devices in his home. Over the course of a decade, that changed rapidly. Middle-class homes first used electric toasters, irons, lights, and telephones. Refrigerators appeared a bit later and spread quickly; when my father was young, his family used an icebox that was supplied with large blocks of ice every few days. The ice was cut out of lakes further north during the winter and stored in special buildings over the summer. Radios appeared in some homes in the early 1920s and spread rapidly to become commonplace in United States homes by 1930. The spread of radios was greatly supported by advances in vacuum tubes. Around 1910, vacuum tubes started to be used to amplify signals and switch currents on and off. Vacuum tubes made possible both the transmission of radio waves and more powerful speakers in home radios and record players.[19] Tubes were the only way to amplify electronic signals until the 1950s. When I was a child, every electronic device in our house relied on vacuum tubes. Not until the early '60s did we have any devices with transistors.

People didn't seem to have much difficulty thinking about pretransistor electronic devices. Taking the conceptual step from a fully mechanical device to an early electronic one was typically achieved by using mechanical analogies for the electronic components. Resistance could be thought of as friction and current as a flowing liquid.[20] Various hydraulic analogies were widely used to help explain circuits. Even the cathode-ray tube that was behind the television screen was often explained in terms of electron "guns" shooting electrons onto a fluorescent screen, like bullets being sprayed to create a pattern on a target. Because no component was much smaller than the eraser on a pencil, you could usually see all the parts of an object and tinker with them. Televisions were more complex electronically than most previous devices, but they did not seem impossible to understand. With a little patience, and a few pointers from an expert, you could learn how they worked and replace faulty components. If you looked at the full circuit diagram of the early black-and-white television set, with a little guidance, you could figure out what all the parts were doing and why.

The very first circuits were easily decipherable blends of electrical and mechanical components. Consider the instructions in an early twentieth-century guide on how to convert a mechanical alarm clock into one that had an enhanced electric bell and illumination of the clock face.

> The alarm clock rests on a shelf, A, which has a piece of metal, B, fastened in such a position that the metal rod C, soldered to the alarm winder, will complete the circuit and ring the bell. The two-point switch D is closed normally at E, but may be closed at F any time desired, thus turning on the small incandescent light G, which illuminates the face of the clock. When the alarm goes off, the bell will continue to ring until the switch is opened.[21]

The conceptual transition to the first electric devices, with components like electromagnets and fuses connected to mechanical parts, was almost seamless. In stark contrast, the transition to transistors soon became overwhelming. A contemporary cell phone may have twenty billion transistors that are best visualized as switches that can be in "on" or "off" positions. The steps from simple on/off switches to the gigantic clusters that enable normal cell phone functions are anything but obvious and seem to evade any simple causal mechanical interpretation. Instead, the assemblies of switches are envisioned as representing the zeros and ones of the lowest level of a symbolic computer code. Eventually all that coding may end up driving a quasi-mechanical operation, such as the vibration of a speaker membrane—but no one sees that membrane as a central part of how cell phones work.

As transistors proliferated madly in integrated circuits, electronic devices migrated from analog operations to digital ones. This shift posed novel challenges to natural human cognition. In the 1830s and 1840s, Ada Lovelace and Charles Babbage, partly inspired by Joseph Marie Jacquard's use of punch cards in looms, started to work on analytical engines and difference engines.[22] When integrated with George Boole's translation of logic into sequences of ones and zeros, these engines ultimately led to the development of all-purpose computers, initially designed as purely mechanical devices and then as electric ones. Before transistors, the most complex electronic digital computers with vacuum tubes were built in the late 1950s and early 1960s. The largest computers at that time had several thousand vacuum tubes that served as switches representing logic gates and could model mathematical operations. Some of these computers could take up a city block. They used large amounts of power and were so expensive that most universities could not afford one. No one imagined that there soon would be several much more powerful ones in a middle-class home.

With the rise of transistors and their ever-shrinking size, within a few decades, a modern cell phone was many thousands of times more capable than the largest computer in 1965. Exotic machines found only in large research centers were now dwarfed in power by handheld devices. This change happened so quickly that my generation went from having a rough idea of how analog devices worked to minimal understandings of digital devices. Most of us were ill-equipped to explain digital technology to the next generation. This was not true when members of my generation were children and were learning—often from parents, older siblings, and friends—how the things around us worked.

Today's children and their children may be learning about the workings of devices in a different manner than previous generations of children did.

As adults, boomers rarely kept up with the new developments and were mostly unable to help prepare later generations for the new architectures of artifacts' interiors. Only a tiny percentage of today's children will ever hear discussions about the mechanisms governing household devices. An even tinier percentage will have an approximate sense of how the parts work individually and together. For the first time in human history, most of the things that we use most of the time have mysterious interiors fathomable by only a few experts. Even those experts won't have complete knowledge; there is simply too much for any one person to keep in mind.

One strategy for dealing with complexity explosions may be to ignore the bottom levels of mechanisms. We could just skip over things like NAND gates and how they give rise to binary codes consisting of strings of ones and zeros and to digital processing. Instead, we might focus on what the larger chips do and use them as our bottom level of explanation. Thus, a cell phone might be described as having an antenna, a display, a keypad or touchscreen, a microphone, a speaker, a battery, a camera, a gyroscope, and a circuit board with a CPU. These hardly exhaust the chip set, but they give an idea of a scale at which quasi-mechanistic explanations might be offered. For example, when asked to explain how a phone takes a photograph and sends it to another phone, an explanation might proceed roughly as follows:

> When you activate the phone's camera, its lens focuses light onto a sensor grid. The grid sends information about the light patterns captured by millions of tiny units, called *pixels*, to an image processor. The processor takes the raw data and formats it into an image that can be viewed on the display screen as the photograph. The photograph is then stored in the phone's memory chip. To send the photograph, the user selects it by using a program that displays it in a larger collection of photographs. The program then uses the phone's central processor to convert the information into a compact form for sending. The phone then sends the photo using radio waves by means of an antenna that broadcasts the photo either directly to another phone or to a cell tower that amplifies the signal and sends it through a network to another phone. That phone's antenna captures the information and shows it on its display.

In one sense, this description, which I generated without referring to sources, is close to tautological. Of course the camera must be turned on and use a lens to capture light and have a sensor surface to register the light; of course the phone must store the information registered in a format that can be used effectively. My "explanation" leaves out all the details of how each step happens and would not enable anyone to build a working device. Yet it is not completely empty, and it does help frame and constrain searches for additional information. It exists at a hybrid level between mechanism and function. Most explanations gloss over many details. When explaining how a manually operated loom works, we don't normally explain how the kind of wood used in shuttles ensures their durability and reliability. An explanation of a loom without the particulars of the wood used for its accessories resembles my explanation of taking and transferring photos with a cell phone. Both go beyond merely discussing function when they describe relations resembling mechanistic transfers of force.

One difference between the loom and phone examples involves multiple realizability. Components such as sensor grids and memory chips can work in many ways and use completely different materials and architectures. In the case of looms, the choices between alternatives are more limited. In addition, the manual loom is a purely physical mechanical chain all the way through. It does not involve the concept of information and its transfer. Even Jacquard's punch card–driven looms did not involve a transfer of information; the cards instead were used to dictate a specific set of steps.

We have progressed from analog, macroscopic, mechanical processes and simple pre-transistor circuits to digital components so small that they are even invisible to microscopes. This kind of change is unprecedented in the history of artifacts. The speed of change over a few decades is also remarkable, as is its rapid spread throughout the world. In roughly fifty years, a complete transformation of parts has taken place (see figure 3.3).

The homogenization of distinct tools into uniform chips does bother us. In June 2024, Apple released a new advertisement for the iPad called "Crush." The video portrayed a massive press squishing beautiful musical instruments, cameras, and many other mechanical devices into a goo that reformed as a gleaming iPad. The advertisement triggered a massive response. People were dismayed at the destruction of distinctive devices that exemplified graceful dances between form and function. Many comments noted that the beauty of device operations—their mechanical instantiations—was also destroyed. As we've seen in discussions of piano players, playing a mechanical piano promotes a kind of awareness of its mechanical underpinnings in ways not found in playing a digital piano. The video was quickly withdrawn.

We have examined how changes from the middle of the twentieth century to the present led to devices with undecipherable components that

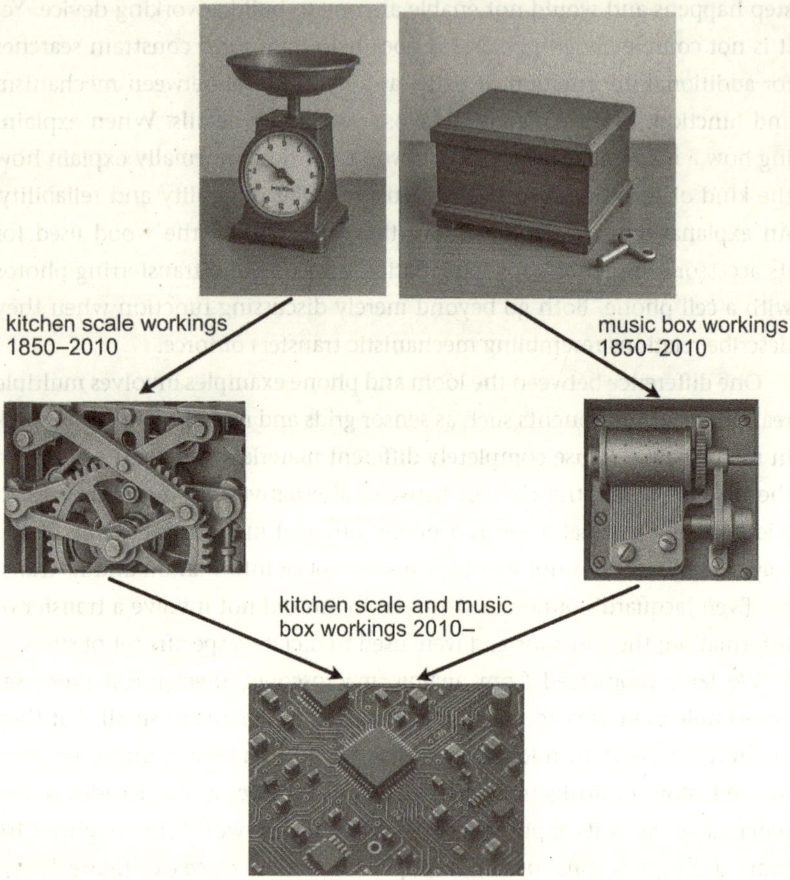

Figure 3.3
Traditionally, most devices causally revealed interactions between parts that were unique to each kind of device. In the past decade or two these parts have become inscrutable circuit boards or chips that are indistinguishable across diverse kinds of devices. *Source*: author.

look the same regardless of their function. We also noted that the first electronic devices were easier to understand, because they could be thought of in quasi-mechanical terms and inserted in causal chains with the other fully mechanical parts of the device. But this story overlooks an important conceptual challenge that may still underlie modern circuits and make them even harder to fully grasp. That challenge is explaining the nature of electromagnetism.

In 1831, Michael Faraday conducted two demonstrations that intimately linked electricity with physical mechanics. A current running "through" a wire caused a magnet to move, and a magnet moving through a coil of wire caused a current to appear. This discovery of electromagnetic induction inspired Faraday to conduct hundreds of experiments on how magnets and currents were related. Throughout his studies, Faraday consistently construed what was happening in terms of physical models such as physical waves and lines of force. He carefully cataloged an interlinked set of regularities that allowed him to describe and predict how the two components interacted. But Faraday was not mathematically sophisticated and did not attempt to express his intuitive model in terms of algebraic equations. That formalization was achieved by the physicist James Clerk Maxwell several years later.

Critically, however, Maxwell said he was able to formalize electromagnetism because of the concrete, visualizable ways in which Faraday talked about his discoveries, such as depicting "lines of force" between magnets and current-carrying wires. Maxwell found such descriptions much more useful than prior mathematical attempts to characterize the same phenomenon.[23] Although he is known for his elegant equations unifying electricity and magnetism, Maxwell repeatedly embraced the use of mechanical models to help inspire those equations.[24]

Thus, a mechanical model mediated the transfer of Faraday's discoveries to Maxwell in a form that he could describe with equations. Faraday's model didn't fully capture every aspect of electromagnetism, but it enabled a revolution in physics and was the foundation of modern electronics. A mechanical model served as a conceptual platform for understanding something that wasn't mechanical at all. This conceptual move also illustrated the potential hazards of using mechanics as a touchstone for making sense of other phenomena. Eventually, Maxwell's equations were only approximations of later, more accurate quantum electrodynamic approaches—approaches that were much harder to envision in terms of simple mechanical models. Even today, however, many people, including science teachers, use physical mechanical models to explain electronic circuits and various phenomena in electromagnetism. Thus, pipes, valves, and a pump might be used to simulate electric currents and resistance.

Maxwell found physical models so useful that he once made a three-dimensional sculpture based on a series of equations developed by one of the pioneers of thermodynamics, Josiah Willard Gibbs. Maxwell's sculpture represented the curves predicted by Gibbs for the freezing and boiling behaviors of an imaginary substance similar to water. Maxwell never met Gibbs, but was so excited by the insights he gained from a physical 3D model that he felt compelled to send it to Gibbs. The model, created in 1874, is still on display in the physics department at Yale, where Gibbs worked for his entire career.[25] It makes visualizing the substance's phase transitions between solid, liquid, and gaseous states easier. For example, by placing straight edges on different parts of the surface, one can see where each substance's unique set of triple points occur, namely the temperatures and pressures at which solid, liquid, and gas states could coexist. Gibbs proved mathematically that all substances must have triple points, and Maxwell's sculpture makes it immediately apparent why that is.

The transition from decipherable mechanical parts to opaque circuit boards and the like is attracting more and more attention. For example, the writer David Berreby recently described how the world had changed for him:

> In the language of 2025, the ancient camera and aged car are "interpretable." That is, my own experience, shaped by widely shared common knowledge, was all I needed to understand them, and other things I used: cameras, typewriters, telephones, tape recorders, clocks, vinyl-disc turntables, stereo speakers and the mimeograph machine. . . . In the 21st century, though, people no longer spend much time in this interpretable world. As I go through my day in the digitally-suffused world, I tell devices to do things for me, and they do, and I have no idea how.[26]

Across most devices and tools, a shift has occurred from the fully mechanical to the electronic. The shift has made many devices more accurate, less expensive, more portable, and more energy efficient. Common artifacts, ranging from clocks to slot machines, have transitioned from mechanical components acting in observable causal chains to integrated circuits in which the inner workings are mysterious to all but a few experts. This shift foreclosed hands-on tinkering and explorations of internal properties; as a result, it may have hobbled the development of explanatory understanding.

* * *

While previous generations used simplified models to understand their tools and surroundings, those convenient fictions helped them navigate reality effectively, despite the incomplete understandings they provided. Critically, our forebears' partial understandings gave them a legitimate sense of connecting with the causal patterns relevant to their goals. In that way, earlier generations developed a strong sense of causality and mechanical understanding through direct interaction with their tools. Today, with mechanisms hidden in circuits, children may have fewer opportunities to develop such cognitive skills. The loss of hands-on learning and the ability to see and understand the inner workings of devices may lead to an impoverished cognitive tool kit.

The disappearance of revealing mechanisms in everyday artifacts hinders our ability to draw analogies and understand new technologies. Children growing up in a world dominated by opaque electronic devices will develop different cognitive skills from previous generations, who learned much through direct mechanical interactions. As technology becomes more advanced and less transparent, a disconnect is growing between users and their understanding of how devices work. If this disconnect is happening, it may be an adaptive pivot to new ways of thinking and learning about artifacts. But even the most positive transitions of this sort can have some drawbacks and unintended consequences. As members of generations that designed and built this new class of things, we are obligated to do the best we can to help future generations adjust to what we have left to them.

These historical precedents of major changes in the parts of artifacts may have affected how people think. Such cases help frame the argument that the most recent change is also consequential. They will also suggest why such changes can go unnoticed by many at the time because of how we all tend to distort our memories of evolving technologies in a larger culture, which leads us to chapter 4.

4 Components and Consequences

On April 27, 1865, the steamboat *Sultana* left New Orleans on a trip up the Mississippi River. After a few early stops, roughly 2,300 passengers were on board, many of whom were recently released prisoners of war from Confederate prisons such as the horrific Andersonville complex. As it approached Memphis, in the middle of the night, the *Sultana* exploded and sank, entombing an estimated 1,800 passengers. The *Sultana* disaster was one of the deadliest maritime events in US history. More people died than on the *Titanic*. Yet this tragedy attracted little attention in the press and is rarely mentioned in histories of the period. The neglect is likely due to multiple causes, such as attention to Lincoln's assassination twelve days earlier and the paucity of written records at the time. But the primary reason may have been the frequency of such explosions on ships since 1800. In the United States alone during that period, several hundred steamboats had boiler explosions resulting in numerous fatalities.[1] While the number dead on the *Sultana* was especially large, it was not news that a steamboat had exploded or that passengers had died.

The explosions reflected the challenges of designing safe steam engines. Advances in the technology of steam engines were hardly the product of one or two people working for a few years. Robert Fulton is often credited as inventing the steamboat at the beginning of the nineteenth century, but others had preceded him (with admittedly less practical models), and Fulton's boats had serious limitations. Major challenges had to be addressed, including improving the metalwork for boilers, creating better safety valves, and controlling water quality. Less obvious, but ultimately more important, was the emergence of the new discipline of thermodynamics. Formal models of gases and liquids enabled engine designers to calculate the effects of engine improvements so as to avoid dangerous pressure and temperature levels in boilers. It took several generations of researchers and engineers across many countries to come up with safe and efficient designs.[2] This is just one of many examples of how the growth of most technologies involves many people over extended periods and does not rely on individuals working in isolation. Even seemingly simple tools, like bows and arrows, took many generations to become efficient implements. In the end, cultural processes can amplify and transform technologies in ways no one person can, but we tend to overlook such processes by exaggerating the transformative contributions of certain individuals. This tendency may be strongest in cultures traditionally known as individualist as opposed to collectivist.[3]

Cumulative Cultural Constructions

We learn from our elders, just as they learned from their elders. While the generational transfer of knowledge has been salient throughout recorded history, more recently a new wave of research has started to explore the process through which knowledge accumulates over successive generations. Academic and laypeople alike have chronically underestimated how much we owe to prior generations and the degree to which artifacts and institutions in our culture support and guide the transmission and growth of knowledge.[4]

The cognitive challenges of learning, preserving, and transmitting cultural knowledge to the next generation may have been critical drivers of human brain evolution.[5] Scholars known as embracing this view of what makes humans exceptionally intelligent emphasize the cognitive demands of curating and transmitting massive amounts of cultural knowledge. Those demands existed even in the most traditional societies. This process of cumulative cultural evolution (CCE) is seen as more central to brain evolution than the challenges individuals faced as lone agents trying to survive in the world.[6] CCE depends on the ability to put information into a language or other symbolic form. The information can then be preserved and transmitted to others. Members of a culture must have a sense of how to convey information to others who may be at a relative disadvantage in terms of basic knowledge, perspective-taking skills, and theories of mind. Cultural transmission requires a sense of what is widely shared knowledge and what is known only by more specialized groups. It also requires an assessment of one's own knowledge deficiencies and how to remedy those deficiencies by relying on specialists.[7]

Artifacts are central to such sociocultural proposals. The cultural transmission of knowledge involves far more than groups of people talking to and learning from each other. Those activities frequently center around artifacts, whether they are cooking implements, agricultural devices, or shelters. Designed and constructed devices not only make life easier for users; they also can convey elaborate information about causal relations. In older traditional cultures, the inspection and use of artifacts revealed directly why they were constructed as they were. Not every subtlety—such as the right way to sharpen a knife blade, or how a certain curve in a plow was needed to turn the soil over to just the right depth for a specific crop—was learned immediately, but those refinements were added on to readily apparent causal patterns relating forms and functions. The economic psychologist Michael Muthukrishna and anthropologist Joseph Henrich describe the process as follows:

> Technologies involving wheels, pulleys, levers, wedges, screws and axles have evolved and spread. Each of these was initially hard to invent (rarely thought of), but once invented, both the application and the underlying concept became easier to learn (from a working exemplar) and re-deployed in novel ways. The wheel, for example, appears to have only been invented once in human history, and only in Eurasia. Initially, wheels were used for carts and pottery, but later the concept was applied to mills, clocks and industrial machines. Similarly, some populations developed technologies, like bows or spring traps, that used elastically-stored energy. But archaeological and ethno-historical evidence indicate that tools involving elastically-stored energy never emerged in Australia, though bows using unfletched arrows did develop in nearby New Guinea. These in turn prevented the invention of related technologies, such as string instruments.[8]

This process may have been augmented by increasing lifespans that enabled grandparents to be reliable retainers and transmitters of cultural insights to the next generation. Advances may also have been accelerated by shifts from hunter-gatherer to agricultural lifestyles that allowed more time for adults to preserve information and transmit it to their children.[9] Children also form their own autonomous peer communities that create, preserve, and transmit cultural information.[10]

Beyond the initial creation of a class of artifacts lay a set of cycles in which the artifacts themselves played a key role in the transmission and elaboration of knowledge. Broad, wide-ranging changes in artifacts influenced the nature of thought in a culture at large; conversely, changes in thought led to artifact innovations. A culture's artifacts loom large in these new cycles. Many artifacts persist across time spans, which enables them to inform future generations. Weapons, vehicles, and tools are common examples.

Over the course of human history, transitions in objects and their components have altered not just the contents of what people normally talk and think about but also how they think. When those changes in thinking occur, they feed back on modifications in artifact designs. In short, there are plausible historical precedents for how a culture's environment, especially its artifacts, can bring about conceptual change.

Not Always Onward and Upward

Technological progress is not an inevitable result of people gathering in groups. Cultures do not march inexorably toward more and more complex artifacts. They can be largely technologically stable for hundreds and even thousands of years; some even undergo technological retreats. One much-discussed case of technology declining over time occurred around 30,000 BCE, when melting ice and rising sea levels destroyed the land bridge between the Australian continent and the island of Tasmania. Inhabitants of Tasmania became isolated from the mainland. Simultaneous climate changes led to food shortages for the Tasmanians. These new hardships in turn caused a decline in population density. Collectively these events led to reduced complexity of tools.[11] The crash in technological complexity may have been inescapable because the reduced population density made it more difficult to transmit cultural knowledge.

This proposal assumes that not enough skilled experts were able to form the kind of robust clusters needed to reliably transmit specialist knowledge. A different proposal has Tasmanians making an active decision to not continue investing time and resources in advanced technology when the situation, because of the changing environment, did not merit that effort.[12] Either way, there was a widespread disappearance of the complex tools used by their ancestors.

A different example of apparent technological decline over successive generations occurred in North America, in the Great Lakes region, from roughly 8000 BCE to 3000 BCE, with the most precipitous drop happening from roughly 4000 to 3000 BCE. The decline involved the making and use of copper tools in forms as varied as knives, awls, and projective points. This technological retreat is puzzling at first, because copper tools have clear functional advantages. But as copper became harder to extract from the earth, and as climate change meant that more effort was needed to procure food, the energetic costs of mining, refining, and smelting copper tools started to exceed their benefits. An additional contributing factor may have been a concurrent rise in societal complexity and rituals involving copper ornaments, which in turn led to less free time and greater scarcity of copper for tools. The knowledge and skills were clearly present, but they evaporated as their manifestation through copper tools melted away.[13]

With the invention of writing and especially of printing, it might seem that technological devolution could no longer occur, as technologies were always recoverable from the written record. But such an assumption neglects knowledge that is transmitted primarily through extensive apprenticeships. In addition, cultures can deliberately devalue technologies and their associated experts. During the Khmer Rouge period in Cambodia, from 1975 to 1979, intellectuals were denounced. People with professional backgrounds, and even those with eyeglasses or hands without callouses reflecting manual work, were transported to rural locations where a great many perished. This policy effectively destroyed many areas of expertise in Cambodia, ranging from several specialties in medicine to much of law. The so-called New Cambodia only lasted four years, between 1975 and 1979, but it marked a radical departure from over two thousand years of Cambodian history.[14] A final, more focused example of technological regression occurred in the Soviet Union, from 1940 to roughly 1960, when Trofim Lysenko banned the discussion and implementation of the principles of modern genetics. Instead he mandated a version of widely disproven Lamarckian views that went against the Darwinian theory of natural selection. His implementation of these discredited science and breeding technologies led to famine and the deaths of millions.[15]

Taken together, these examples from prehistory and the relatively recent past demonstrate that technological complexity does not ratchet ever upward as a consequence of cultural cumulation. One doesn't need to be a dystopian futurist to acknowledge that future progress is not inevitable. The appropriate conditions must exist.

Groups of Experts

Intrinsic to the idea of CCE is the rise of groups of experts who occupy valued niches in communities of knowledge. A recent body of research on divisions of cognitive labor and communities of knowledge is revealing the complexities involved in successfully navigating the multilayered terrains of expertise in a culture.[16] Outsourcing knowledge to experts is often construed as a contemporary phenomenon, but aspects of this process have been inherent in cultures of any complexity for millennia. We will ask in later chapters whether modern technology is leading us to new and potentially more hazardous forms of outsourcing.

While the advantages of specialization are well known, they can overshadow another equally important factor: the need for shared causal understandings that enable interactions between different experts and members of diverse groups. In any community with divided expertise, interactions among different groups will be more efficient when everyone has a certain level of shared understanding. The precise level varies with the task, but some degree of shared knowledge is needed to facilitate communication among a group's members. Some researchers have recently argued for placing special value in shared technical or mechanistic knowledge.[17] If so, helping children build on their early interests in mechanisms could bring future benefits for a community. This also involves helping children learn more about the limits of their own technical knowledge and how to consult others.

In short, humans have uniquely illuminating insights by exploiting information that is captured and transmitted through groups that make up their culture. Culture has expanded the minds of individuals over successive generations and may have influenced the last wave of brain evolution in humans. Artifacts have been central to that progression, so we now turn to consider three dramatic artifact changes at different points in the past. We will ask how such changes influenced how people think, understand, and explain. In addition, we want to know if the strongest effects might occur in children. Early experiences with the material creations of a culture may set up essential cognitive foundations that enable the development of more refined intuitions about how things work.

The Critical Role of Physical Artifacts

Discussions of CCE often focus on culture in terms of people, their interactions and roles, and what they say. This can result in relegating a culture's artifacts to supporting roles as props for social interactions. But the artifacts themselves serve as powerful influences on thought and innovation. Artifacts can mediate cultural transmission both as containers of rich knowledge and as teaching vehicles.

The transformative importance of artifacts is apparent when an artifact transforms an individual's intellectual life. One celebrated case occurred when a particular instrument helped a young woman become a leading scientist against all odds. Maggie Aderin-Pocock was born in 1968 in London after her parents immigrated from Nigeria. Her parents divorced when she was young; that separation, plus other events, resulted in her attending thirteen different schools before her eighteenth birthday. Not surprisingly, the constant changes led her to became alienated from school settings. Her prospects seemed dim until a particular artifact captured her attention: the telescope.

Aderin-Pocock had an interest in the moon and space ever since early childhood, but this interest took off when, as a fourteen-year-old, she signed up for a class that involved making her own telescope. She described the telescope experience as follows: "For me this was the first instrument I made and it was fantastic because I made it with my own hands, and it got me closer to the stars I loved. . . . I wanted to understand the stars, so I needed a telescope."[18] Aderin-Pocock has described how her telescope enabled her to see new details in the moon and how these insights inspired her to make better telescopes. This passion led to an immersion in science-related topics, and her grades soared. She earned an undergraduate degree in physics from the elite Imperial College London and went on to receive a doctorate in mechanical engineering from Imperial as well. Her experiences making and tinkering with telescopes turbocharged her intrinsic interest in space and related technologies, and she went on to become an internationally celebrated scientist. She designed numerous leading-edge instruments, including several satellite imaging devices, multiple instruments for tracking climate change, and handheld landmine detectors.[19] Without her transformative early experiences building telescopes, none of this might have happened.

Prehistoric Tool Transitions

At certain points in the history of humans, the parts constituting artifacts have undergone radical changes. Of particular interest are parts changes that affected how humans thought about a wide range of categories. We start in prehistoric time, without written languages, for which artifacts are examined by archeologists to make inferences about effects on human thought.

Roughly two hundred thousand years ago, humans started making tools containing parts with different functions. The first of these composite tools often used hafting technologies. Hafting is the binding of one rigid object to another with a flexible material. One of the earliest examples is the use of plant fibers to bind sharpened rocks to wooden shafts.[20] Hafting emerged independently in different groups, some of whom preceded modern humans.[21] With the emergence of modern human brains, hafted spears, axes, and pounders were soon followed by a rapid proliferation of other composite tools such as scrapers, throwing sticks, and travois. Because this second revolution occurred after prehistoric human brains were thought to be the same as ours, the emergence of diverse composite tools has been interpreted as occurring primarily through the cumulative cultural evolution of cognitive capacities. While some now see CCE as exerting selective pressure for larger brains, as more people lived near each other and lived longer, the growth of cultures alone may have been adequate for supporting technological transformations.[22] As cultures advanced, sharing and transmission of knowledge across generations became easier and enabled each generation to more effectively build on its predecessors' accomplishments.

Clocks

In the early 1300s, the perceived roles of parts started to shift following innovations in clock design. Clocks were able to run with improved accuracy for many hours without external sources of power. Instead they could rely on internal stores of energy such as suspended weights and compressed springs. This technological advance arose from the development of new kinds of escapements that allowed evenly paced releases of stored energy. Energy storage advanced with the development of more efficient metal springs.[23] Clocks became more portable and now seemed capable of self-contained behaviors. The rise of such clocks inspired comparisons to animals, plants, and other natural entities outside the realm of biology.[24] The extension of clock-like mechanisms beyond clocks and related devices accelerated with improved designs linking a steady oscillation frequency (weighted balance wheels and hairsprings) to escapements.[25]

Discussions of the natural world in terms of clock metaphors were largely shallow comparisons until the seventeenth century. Animals might be compared to clocks because they had hidden natures that helped explain what they did, but they were not seen as operating just like clocks or other machines. They continued to be seen as organic creatures with unique properties, such as essences and vital forces. However, in the early 1600s, comparisons deepened and much of the world, living and nonliving, was described as operating with clocklike components in elaborate mechanistic chains.[26] Descartes was a well-known early proponent of this view. He described animals with interiors much like those of clocks, but much smaller and invisible to the human eye.[27] Comparisons to clocks may have taken off in the seventeenth century because clocks were no longer found mostly in churches and public buildings; they became more common in homes. Comparisons to nature may have seemed more compelling as clocks were more intricate and self-sustaining for longer periods. But the surge of interest may have also been due to the influence of a few especially compelling individuals, such as Descartes.

For the mathematically inclined, such as Isaac Newton, clocks were examples of lawlike, elegant physical mechanics that could be neatly captured by mathematical laws. Newton and many others found this formalization appealing as they assumed that mathematics was one of the purest demonstrations of God's creative abilities. The math hidden in all of nature supported their version of God as solely responsible for all of nature's inherent structures.[28]

The tendency to think about organisms as machines was further nurtured by the appearance of automata in the early eighteenth century. Automata seemed to vividly demonstrate the plausibility of equating machines with the internal operations of animals and even humans. Two celebrated examples were designed and built by the French engineer Jacques de Vaucanson in 1737 and 1739. The Flute Player, a mechanical humanoid, played a dozen songs by moving air through tubes and thereby "breathing." It simulated a human skill and fascinated the public at exhibitions. The public became even more enamored with the "digesting duck" (figure 4.1). This duck was described as simulating the physiological process of digestion.[29] It would bend its neck to get small bits of grain, move its wings in a manner suggesting contentment, and then seem to produce appropriately digested waste. It also quacked, appeared to drink water, and swallowed. Years later it was discovered that the duck did not digest the grain: Vaucanson had hidden a small sack of green-colored flour in a bag that, when squeezed by the mechanical parts, excreted duck waste at appropriate moments.

The duck astounded people from all walks of life. Voltaire repeatedly raved about it, at one point exuberantly stating that "without the voice of

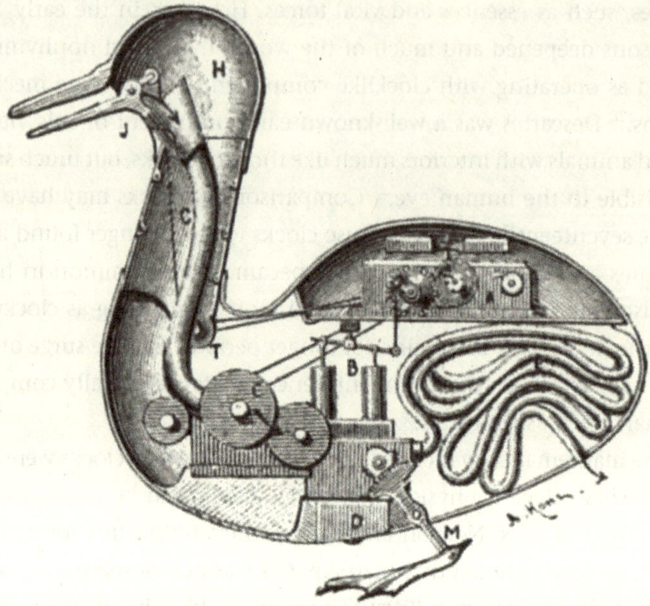

INTERIOR OF VAUCANSON'S AUTOMATIC DUCK.

A, clockwork; *B*, pump; *C*, mill for grinding grain; *F*, intestinal tube; *J*, bill; *H*, head; *M*, feet.

Figure 4.1
An artist's interpretation of Vaucanson's digesting duck. *Source*: *Scientific American*, 1899 (Vol. 45, p. 43).

de la Maure, & the duck of Vaucanson, you would have nothing that would remind us of the glory of France."[30] (La Maure was the most celebrated opera singer of that time). The special fascination with the duck may have reflected an appreciation of how a machine could duplicate even the mysterious microscopic processes of digestion. The duck continued to live on in various novels, most notably in Thomas Pynchon's *Mason and Dixon*, where it had been supposedly enhanced by Vaucanson in ways that led to self-awareness.[31] Across many of these fictional accounts, writers seemed to quickly leap from automated self-contained action to the ability to have goals and desires. The wider public's giddy fascination with automata illustrated, for perhaps the first time in history on a large scale, how lay people were profoundly interested in inner workings and by the possibility that all of nature was composed of "clever" machines.

Interchangeable Inspirations

Around 1800, artifacts transformed from bespoke creations made of unique parts to mass-produced copies with interchangeable parts. Before 1800, a soldier on a battlefield with a broken rifle had to find a gunsmith to make custom replacement parts. Every trigger assembly was unique to each gun. Interchangeable parts made for faster repairs, lower costs, and mass production. These new standardized devices appeared first in the military, but the revolution soon spread to sewing machines, bicycles, and virtually all other commonly used appliances, tools, and vehicles. The shift did not occur earlier because it depended on the emergence of precision machine tools. Without tools that could reliably and accurately reproduce metal components such as screws, bolts, and gears, interchangeability was impossible. The difficulty of making precision tools led to a grand deception of the US Congress in 1801. Eli Whitney dazzled members of Congress with a demonstration of allegedly interchangeable parts for a set of rifles whose parts were shown on a table. His demonstration was a sham, however, as he had not been able to master the precision machining needed for the parts.[32] Instead he performed a shell game in which the same rifle and parts appeared and disappeared to give the impression of being multiple rifles. Whitney received the government contract but failed to deliver the promised rifles. Others did achieve that precision and mass production a few years later.

The transition to standardized parts offered new ways of thinking about devices and how to modify them. Identical components support counterfactual thought, making it easier to consider what-if scenarios regarding new designs. For example, if a company made rifles with .38 caliber barrels and a competitor made rifles with larger .44 barrels, the presence of two standardized sizes would have promoted questions about what would happen if both companies produced barrels of the .44 size. Predicting the effects of part modifications became easier. When every part was customized for a unique device, changes in part design were less predictable. For example, when cartridges were developed for breech-loading rifles, the thickness of the brass and the fit with the chamber were critical and variations could have catastrophic consequences. Standardization also enabled rapid convergence on more cost-effective and safer designs.

The philosopher of science William Wimsatt suggested that interchangeable parts often became generative conceptual building blocks.[33] Those blocks could then be exported to other devices as stable functional units. When a component became widely used across devices, its internal mechanism could afford to be a black box as its function moved to center stage. A multipart standardized component that served as a universal timer in everything from washing machines to toasters might only need to be understood in terms of its inputs and outputs. Black boxing may have thereby supported the increased transfer of a functional component across diverse devices even as it neglected internal mechanisms. Wimsatt describes several ways that interchangeable parts may have had cognitive consequences over the decades when such parts came to dominate almost all manufactured goods. This transition, which is largely unknown to most people today, represented a dramatic shift in how almost all devices and tools were made—a shift that influenced the nature of future designs and innovations as well as how we understand the artifacts around us.

Wimsatt illustrates the potential conceptual depth of the parts shift in his description of Herbert Herkimer's 1952 book *Engineers' Illustrated Thesaurus*. Although it is rarely cited, this thesaurus of devices is an impressive conceptual guide. Herkimer explicitly embraced Peter Roget's goal of organizing all knowledge into useful conceptual clusters, but Herkimer focused on devices instead of knowledge.[34] He started with high-level categories such as fasteners, adjusting devices, supports, and mechanical movements. Each of these was further divided into smaller categories. Kinds of mechanical movements included levers, connecting rods, block and tackle, differential winches, tappets, cams, rack and pinion, toothed gearing, epicyclic gearing, pawl and ratchet, reciprocating motion, differential gearing, pitch chain drives, friction clutches, shaft clutches, angular clutches, and springs. These categories and accompanying line-drawn illustrations enabled readers to infer the abstract relations common to members of each category and to recognize novel instances. Herkimer's examples were based on manufactured objects that were interchangeable parts, contained interchangeable parts, or both. He grounded his intuitions by appealing to the classic proposal that all machines, however complicated, can be reduced to six basic types: levers, wheels, axles, pulleys, inclined planes/wedges, and screws. This ancient claim works well even today.[35]

While Wimsatt discussed black boxing and a greater emphasis on functions as possible consequences of the shift to standardized parts, a more subtle thematic change may have also occurred in the ensuing decades that led to a reframing of how we think of all tools and devices. Prior to the shift, parts, and by extension entire objects, were more likely to be seen as individual adaptations to unique environments and situations. Rifles might be tailored to the user's anatomy and local ammunition. Most wagons were customized to a user's needs (e.g., fast vs. rugged transport, things to be carried, terrain, distance traveled). When making a new tool or device, the maker would want to know about the user, the context, and how the object would be used. In that sense, many artifacts were seen as individuals whose properties were dependent on users and the local environment. They were highly specific adaptations.

As objects acquired interchangeable parts, the benefits of lower costs and easier repair were obvious, and the change rapidly spread. Devices became conceptualized as standardized "models" and not as individuals. A printing press might be identical to all other printing presses of the same kind. Things no longer seemed like adaptations that were deeply embedded in their environments. Instead they were construed as freestanding entities, remaining constant across environments.

This change incurred cognitive costs as well. People became less aware of how tools originated and why they were designed for a particular context. In each setting, many tool properties were optimized to achieve goals and suit situation-specific physical constraints. Missing those details may have reduced mechanistic insights. For example, noticing how the thickness of a rod varied about a fulcrum might guide learning about the roles of levers as parts of devices and suggest reasonable types of innovations. These bespoke variations highlighted causal relations in ways that standardized parts could not.

These brief sketches of prehistoric multipart devices, clocks, and devices with interchangeable parts hardly exhaust the ways that new artifacts impact cultures. Devices such as the printing press, the steam engine, and the cotton gin clearly had massive effects, but these effects were arguably less strong in terms of how we think and talk about the world around us. The three cases described here were chosen because they are closely related to the shift to uninterpretable parts and to where childhood experiences may matter most.

Toys and Play

Toys are important to the growth of technology; they are more than simply fanciful imitations of adults' artifacts. Toys are often small working models of life-sized counterparts. They occur in all cultures as objects ranging from miniature weapons to cooking utensils. Toys have never simply been things to entertain children. They help transmit cultural ideas and practices to the next generation. In many anthropological studies, children's play with toys is described as crucial to learning about technologies that children must master by the time they are adults. For example, playing with toy bows and arrows teaches children about the effects of using different materials, dimensions, and tensions. Most of the time, this learning is through free play and not through pedagogical demonstrations by adults or older children. As one overview of fifty-four different hunter-gatherer cultures put it, toys themselves were the teachers.[36]

Do contemporary toys serve the same roles of enhancing cultural cognition and providing children with important knowledge for adult life? The question has not been extensively examined but deserves a closer look given changes in the types and uses of common toys. Some of the most popular toys in 1960 included the Slinky, the Easy-Bake Oven, Barbie dolls, Chatty Cathy dolls, and Matchbox cars. Some of the most popular toys in 2022 included an electronic dance mat, Lego Gear Bots, and remote-controlled stunt cars. These toys have more electronics than the earlier ones. Even Chatty Cathy was surprisingly mechanical. To make the doll speak, a ring attached to a string going into the doll's belly was pulled. The string was connected to a phonograph record inside the doll, behind its abdomen. The record was driven by a metal spring that was wound up by pulling the string. All this occurred without any batteries. When toys broke in 1960, it was usually easy to see what went wrong and how to fix it.

Young children today spend most of their time with toys that reveal little about their inner workings. In contrast, children fifty years earlier had extensive exposure to toys with interpretable workings. These different experiences are another way in which children now might have fewer rewarding insights about underlying mechanisms. Perhaps the loss of information about inner workings is more than compensated by a new class of insights—but no evidence to date supports that view. In contrast, one review of the impact of digital toys in comparison to traditional ones found no systematic advantages and raised concerns about how digital toys can be overstimulating, socially isolating, and less likely to foster free play.[37]

Some modern toys diabolically work against mechanistic intuitions. When our preschool-age twin granddaughters received a toy saxophone, the cognitive chaos began. Their parents initially noticed a confusing mechanism in which blowing into the mouthpiece activated a sensor that spontaneously changed musical tones. The keys were even more mechanically misleading. The player could hold down only one key at a time, and the lower one moved down the row of keys, the higher the sound. This contradicts the physics of real saxophones: moving down the stem of a saxophone lowers the tone as the enclosed space is larger and reverberates at a lower frequency. It is difficult to imagine creating a worse toy for building intuitions about how sounds of different frequencies arise.

Informal conversations with young children about contemporary toys and their earlier versions can reveal striking contrasts in both the conversations and their ensuing actions. When another grandchild was six, we asked her first how a present-day music-making toy worked and then how a 1971 version worked. The modern version consists of a box with a slot into which various cards are inserted. After a card is inserted, the box plays a song specific to that card. The 1971 version is a "record player" with plastic disks that can be placed on a rotating turntable to make specific songs occur. The modern machine is an all-digital device; each card contains a song that is transmitted to an electronic speaker when the card is placed in the slot. The record player, in contrast, has no electronics but contains a comb with metal strips in the player's arm. Each disk has bumps on its surface that pluck metal strips of different lengths, corresponding to different notes.

The modern music box is made more complex because it plays songs in two different ways. Some cards contain digital content that is transferred wirelessly to the box. Other cards contain a code that enables the box to play songs by streaming them from a Wi-Fi network. When we asked our granddaughter how the box worked, she paused and asked us what we meant. We then asked how each card made it play its own song. She indicated that the box had a "sensor" that could tell which card it was. We then asked how that worked and she grinned and said it was magical. Beyond that she shrugged and indicated little desire to dig deeper into mechanisms. Her responses may seem childlike until we realize that most adults' answers are similar. With two completely different mechanisms underlying a card's activation of songs, it's no surprise that the boxes don't easily reveal how they work.

In contrast, the record player immediately suggests mechanistic chains by the different bump patterns on the disks and by observing that the songs slow down if one rests a finger on the disk or the wind-up spring runs down. Our granddaughter found it fascinating and soon figured out how it worked, with a little guidance from her parents.

One final set of toys is less common in most children's lives today: kits for children to assemble working devices from nearly "raw" materials. Two of the most popular kits in the 1950s and 1960s involved building an electric motor and a crystal radio. The parts were little more than wires and a few crude parts that could conceivably be scavenged from common materials. Versions of the motor that I assembled in 1961 can still be found for sale online. These kits were sold by the GW Moore company from the mid-1930s to well into the 1960s, with hardly any modifications over that period. The motor was built on top of the cardboard box it came in.

The motor kit contained wires, small permanent magnets, and various pieces of metal, most of which could be bent from straight pieces. I can still remember the thrill when I first connected the motor to a battery. It immediately started spinning. These kits were widely distributed to public and private schools across the United States and constitute a shared memory among many of my generation. Compare those kits to modern versions, which are rarely found in schools but can be purchased online. The simplest one is refreshingly like the 1930s version, with a bit more distance from raw materials. It is harder to visualize how to make it from raw materials, with its cast plastic components and premade wire connectors and battery holder. More concerning is its dismal popularity as evidenced by low ratings and sales. It was ranked number 438,147 on Amazon in terms of popularity in toys and games and in the top thousand in the category "Educational Science Kits," with an average review score of 2.9 out of 5. As of April 5, 2024, the top-ranking item in toys and games went to a GI action figure. In educational science kits, the top item was a kit for growing your own butterfly consisting of a pop-up mesh container, a plastic dish called the feeder, and a voucher for ordering baby caterpillars. Tinkering and exploration do not seem to be viable options here.

More complex build-your-own kits are available on Amazon, but these typically have elaborate circuit boards among their basic parts. One popular example has children build their own robot. The kit has 190 pieces. A science experiment kit contains solar panels and solid-state circuit boards. Consumer reviews suggest that the only truly mechanical parts are poorly designed fragile plastic ones that frequently break. The primary emphasis of the kit as a whole is on the unexplained solar panel and the sealed motor. A migration has also occurred from mechanistic descriptions of parts to descriptions that are mostly functional (e.g., one description discusses the solar panel as designed for capturing the sun's energy and turning it into electricity that powers the robot). Solar panels pose an explanatory dilemma, as any understanding of how they work requires substantial knowledge of physics.

Some studies suggest that traditional toys promote richer language and communication episodes with parents, while more modern toys tend to dominate the situation and result in children becoming passive observational participants.[38] But even if such effects are repeatedly found, they may not necessarily result in fewer insights into how things work. Moreover, if correlations between toy types and mechanical knowledge are found, they might arise from lower-quality interactions with adults and peers rather than from the toys directly. All these alternative explanations must be systematically ruled out. That said, we cannot ignore the worrying possibility that most modern toys may represent regressions in terms of providing supportive frameworks for building understandings of how things work.

Boomers to Zoomers and Language

Have part changes since the 1950s and 1960s had systemic influences on thinking? If no mechanistic information can be gleaned from inspecting insides, people could start talking and thinking less mechanistically about objects and their components. To make generational comparisons easier, here we'll mostly compare *boomers* (baby boomers, born between 1946 and 1964), and *zoomers* (gen Z, born between 1997 and 2010). In most instances, the arguments hold for adjacent generations.

Natural language use in different time periods provides clues as to the content of everyday thoughts about devices. We can examine whether mentions of mechanical parts that have been equally common in devices in 1950 and 2023 are decreasing in frequency over time. For example, several key parts of car suspensions have not changed over the decades. Coil springs, shock absorbers, and tie rods are just as common in 2023 Teslas as in 1950 Fords. Yet in databases that can be queried for frequency of word use over the years, downward shifts do seem to repeatedly occur. Google Ngram tallies the frequency of words and word strings for every year in books published between 1800 and 2022. The left panel of figure 4.2 shows graphs of frequencies of mention from 1950 to 2022 for the terms *coil springs*, *tie rods*, and *shock absorbers*.

Tie rods were mentioned ten times more frequently in 1950 than today, shock absorbers three times more frequently, and coil springs seven times more frequently. Contrast these mechanical car parts with other parts of cars that are easily visible but serve no mechanical role: windshield wipers, glove compartments, and side mirrors. The frequencies stay flat or rise (figure 4.2, right panel). The use frequency of terms for internal combustion parts drop

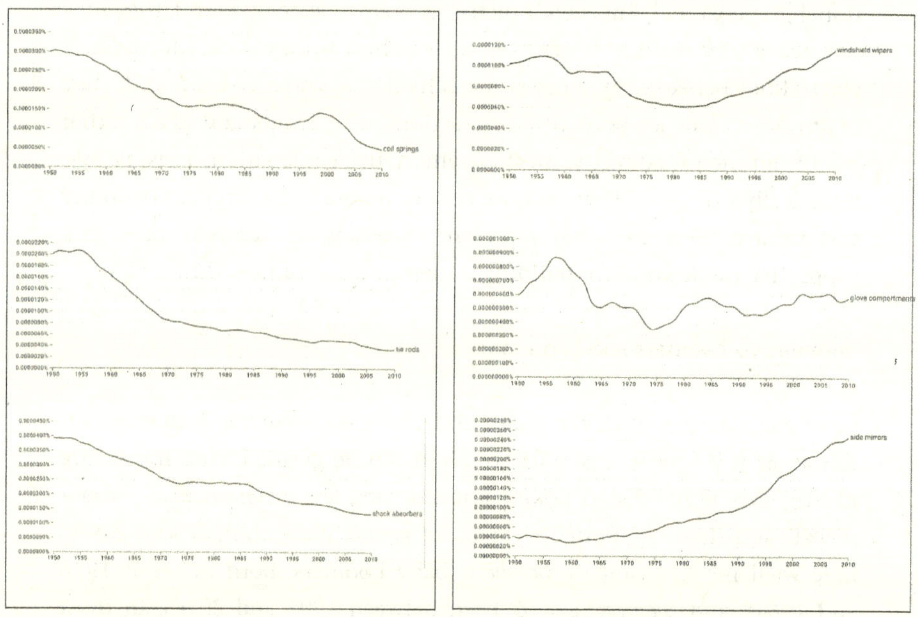

Mechanical Parts Nonmechanical Parts

Figure 4.2

Left three graphs show declines from 1950 to 2010 in usage of the words *coil springs*, *tie rods*, and *shock absorbers*. Right three graphs show usage over the same period of the words *windshield wipers*, *glove compartments*, and *side mirrors*. Units on *y* axes vary so as to more easily visualize relative declines. *Source*: author.

dramatically as well. (Because of the recent rise of electric cars, terms for internal combustion parts are not shown here, even as the relatively small percentage of electric cars would not predict the size of the drops.)

The same pattern generalizes to the working parts of other devices that have remained relatively unchanged over the decades. Over the past seventy years, household refrigerators, for example, have consistently contained compressors, expansion valves, and cooling coils. Yet those parts are mentioned one-tenth as often today as in 1950. The decline of words referring to mechanical components in common discourse is not mirrored in the language of experts. For example, the percent of phrases in all text strings in Ngram for *glove compartment* quickly outpaces that for *coil springs* between 1950 and 2020, but the number of times those phrases occur in Google Scholar shows the opposite pattern.

Some nonmechanical parts have existed in most refrigerators since 1950, such as ice cube trays, crisper drawers, and freezer doors. No Ngram graphs for any of these show a downward trend over the years. People talk about refrigerators just as much or more than they did seventy years ago; they just don't talk about mechanisms. They talk more about "features." Similar declines occur for the names of common mechanical parts found in many devices, such as lock washers, bushings, machine screws, retainer rings, spindles, rocker arms, worm gears, hanger arms, helical gears, and O-rings. Every item shows declines in usage since the 1950s even though all of them are just as ubiquitous in devices today. Whether it's a Roomba or a hoverboard, the latest things that move about have similar mechanical components. The data for the term *O-ring* validates the sensitivity of the Ngram measure by declining continuously except for an upward blip around 1986, which is when the space shuttle Challenger exploded because of a notorious O-ring failure.

More exhaustively, consider *every* item listed by Wikipedia as a "machine element" on September 14, 2022. For just those that had enough numbers to register on Ngram and did not stay flat in frequency, thirty-three showed drops and five showed increases. Two of those increases (for *keys* and *nuts*) are explained by word ambiguities. The average increases were also much smaller than the average drops.

The most dramatic case of large declines in language use about mechanical parts occurs in Herkimer's list of types of mechanical movements. Omitting terms that appeared so infrequently that Google Ngram does not plot them, the terms for *every one* of Herkimer's items (levers, connecting rods, block and tackle, differential winches, tappets, cams, rack and pinion, toothed gearing, epicyclic gearing, pawl and ratchet, reciprocating motion, differential gearing, pitch chain drives, friction clutches, shaft clutches, angular clutches, and springs) show declines in use. Moreover, in almost all cases, the drops are dramatic. Given that Herkimer selected these types of movements because they were conceptually most useful for his thesaurus, it is impressive how consistently less they have all been written about from the 1950s to the present. People are not talking less about artifacts in general. Discussions of nonmechanical parts and properties generally stay constant or increase. In addition, most of Herkimer's items have not been disappearing from devices. Almost any device that moves in any way still must employ these parts, as there are few alternatives.

The discussions of mechanical parts have not been displaced by language about electrical parts. If we examine electrical parts that have been present for several decades (e.g., capacitors, resistors, transistors, and transformers) their Ngram charts all show drops over the same period. Terms for common internal components in computers over the years, such as *CPU* (central processing unit), *RAM* (random access memory), and *motherboard* have declined in usage. This neglect of basic causal underpinnings of computers has been noted repeatedly by the shrinking group of experts who do know about such parts. Even majors in the discipline no longer have insight into basic mechanisms such as how simple configurations of switches can model logical operations such as "and" and "or." The greater public is even more removed from any sense of how computers give rise to their functions.

Google's NGram data is not fully representative of natural language use. It draws from only the text that occurs in books. Ideally we would like to analyze annual frequencies of word use in all other text outlets as well as in spoken language. Unfortunately, comparable datasets showing word use numbers in these other formats for every year since 1950 are not easily available. The full spread of books covered by Ngram is large and provides some reassurance that the differences over the years are likely to be reflected in daily discourse. While only roughly 10–15 percent of those books are fiction, it is reassuring to see that similar decreases of mechanical parts terms and no decreases of nonmechanical parts terms are found when only works of fiction are examined. The variability is higher because of smaller numbers of instances, but the overall contrasts are strikingly similar.

Modern Electronics

The mechanisms involved in electronic circuits are different from those in mechanical movements. We do still see drops in the frequency of mentions of inner mechanisms for electrical devices and their components, such as resistors and capacitors or the basic architectural features of computers. This decline may occur because older circuits tended to reveal their mechanisms. In early circuits you could see each component and visualize how they interacted. Those interactions are invisible in modern devices. This is not because people in 1920 could really "see" dynamic causal interactions among electrical components, but they were able to employ mechanical metaphors, especially of fluid dynamics, to visualize causal relations and make predictions. The total number of electrical components was small (say under a dozen) and visible. But then households were flooded with transistor-driven devices in the 1960s. These soon evolved into solid-state circuits and chips with thousands, millions, or billions of transistors. Discussions of single components and their interactions disappeared. The meaning of components and their interactions switched from mechanical causal chains to computational routines and symbolic operations.

This change from the quasi-mechanical to the computational was especially vivid to children in my generation, who saw a transition in how television remote controls worked. Our first television remote had buttons that, when pressed firmly, caused a metal hammer to strike one of four metal bars. Those bars varied in dimensions much like those in a tuning fork, except the sounds produced were of higher frequencies. Each bar triggered a microphone in the television that was tuned to the frequency for that bar. The activated microphone then completed a circuit that caused a motor to turn a channel or volume knob up or down. You could hear the metal ding and then the whirring of a motor in the TV as the channel or volume knob turned in the desired direction. Yet, just a few years later, television remotes used electromagnetic signals, and their principles of operation became undecipherable.

Beyond Artifacts

Could mechanism neglect also spread to some natural kinds, such as mechanistic components in biology? Returning to Google Ngram, we do find that the names of many parts of eukaryotic cell mechanisms, such as *vacuole*, *mitochondria*, and *nucleus*, have appeared less frequently since the 1950s, while mention of superficial biological parts, such as freckles, birthmarks, and eyebrows have not, and in many cases have soared. While the decline could have occurred for other reasons, we can imagine how the decline in decipherable artifact parts might have had consequences for other domains. Mechanical expressions abound in many descriptions of molecular processes.

We have seen that a surge of descriptions of living things as clockwork devices appeared in the late seventeenth century in the writings of Descartes and others. As clocks became more commonplace during that time, changes in those artifacts may have influenced thought about the living world. There are many pitfalls to excessively thinking of the living world as machines, but for laypeople, mechanizing the world to a fault may have been more useful than relying on a vague vitalism.[39] The ease with which laypeople could visualize biological components in mechanistic terms might well be a function of their experiences with easily decipherable artifacts. Those with greater experience are likely to also be more able to see the limitations of mechanical descriptions in biology.

All these examples of potential changes in the contents of thought in recent generations refer to correlations, not experimental manipulations of exposure to mechanisms over the decades. Alternative causes can be offered for each case. But an overall pattern of shifts from mentioning mechanistic parts to discussing superficial features and functions is difficult to ignore.

Starting Again from Scratch: The Connecticut Yankee Conceit

We can better appreciate just how much our knowledge builds on information accumulated by earlier generations when we consider cases of individuals attempting to create artifacts from scratch. If divisions of cognitive labor are powerful engines of technological change, what happens to humans in situations where that division of labor no longer exists? This scenario has been repeatedly depicted in novels where a person travels back in time to an earlier era. The traveler has the widely shared knowledge of the present day while all the people they encounter have the shared knowledge of that time in the past. The visitor's successes at recreating all or some of the present are the central theme.

These time travel tales tend to grossly underestimate the challenges the traveler would face. For example, in Mark Twain's *A Connecticut Yankee in*

King Arthur's Court, a mechanic named Hank, who works in a nineteenth-century arms factory in Connecticut (Twain's present), is transported into the Arthurian era (528 CE). Hank immediately awes the people of that time by predicting an eclipse. He then goes on to easily create devices from his era, such as modern guns and bicycles.[40] In the end, Hank's technological improvements are rejected by the powerful church hierarchy, which sets out to destroy him. Hank escapes by returning to his own era.

Twain's primary purpose in the book was to make fun of religious hierarchies and romantic notions of the Middle Ages, not to consider the plausibility of Hank's technical creations. Nonetheless, Twain ignored seemingly obvious difficulties of creating modern artifacts so easily and rapidly. For example, manufacturing many bicycles would have required making interchangeable parts. The existence of such parts depended on a level of precise machine tooling that was not yet available, which would also have been required for the manufacturing of numerous revolvers.

A more optimistic description of an individual bringing technology to the past occurs in the Conrad Stargard novels by Leo Frankowski, also known as *The Cross-Time Engineer* series. A Polish army engineer is transported back to 1231 CE and soon helps accelerate his country's technology to safeguard it from a Mongol invasion. He knows from his history studies that the invasion will come in about ten years. He succeeds and goes on to create a glorious future for Poland. In yet another popular novel, L. Sprague de Camp's *Lest Darkness Fall*, an archeologist, Martin Padway, is transported back to Rome in 535 CE. Padway introduces several new technologies and creates a better future for the members of that society.

A strong contrast to these positive outcomes is found in Poul Anderson's short story "The Man Who Came Early." An engineer goes back to medieval Iceland and is a complete failure. He cannot adjust to either the existing technology or the culture. Anderson clearly was responding to how easily the protagonists in other time travel stories reintroduce technology.

Most of the authors of these tales were engineers and scientists, as are most of the protagonists. People with such backgrounds might be expected to have more realistic senses of how technology and science rely on communities of experts. Instead they seemed to have especially inflated views of what solo experts could achieve—an illusion that often recurs when experts are asked how much they relied on the skills of others to make new artifacts.

Dystopian Regressions

A different approach to reinventing technology comes in dystopian speculations about future disaster scenarios. These tend to be more detailed and realistic reflections on what one would have to know to rebuild the modern world. A particularly memorable example of the challenges encountered in building things from scratch is recounted in Thomas Thwaites's book *The Toaster Project: Or a Heroic Attempt to Build a Simple Electric Appliance from Scratch*. Thwaites tried to make a working toaster from raw materials found in the wild. This is no easy feat, and Thwaites gleefully documents his rocky journey. For example, many parts of a toaster are metal. To procure those metals, Thwaites goes back to the original natural deposits in the earth's crust. Just this initial step soon overwhelms him as a solo agent. He must consult with experts to learn what locations on the earth contain likely veins of ore that might be extracted and refined. Without that help he would have wandered aimlessly through the countryside, sifting through soil and rocks that contained scant traces (if any) of the required copper, iron, and nickel.

Thwaites soon cheats by consulting with a metallurgy expert. The expert indicates that even if Thwaites went to the highest-yielding mines in the world, he'd still probably need to sift through a ton of raw material containing each metal to get enough to make the metal parts of a single toaster. Thwaites plans to also find the raw materials for the toaster's plastic parts, but soon learns that creating plastic from raw materials involves complicated processing of crude oil, which explains why plastics arrived technologically much later than metal artifacts. Thwaites is undeterred, however, and later unsuccessfully tries to synthesize plastic as well.[41]

Thwaites acts on the expert's information about good locations for iron ore. He travels north from London to a mine that has yielded high-quality iron for centuries. Again, Thwaites must cheat. He cannot extract the necessary rocks from the mine on his own, and he doesn't even know where to start digging. He commandeers a pickax but soon discovers that it is utterly inadequate. He ends up taking another shortcut when a benevolent veteran miner simply gives him forty kilograms of especially high-yield ore, which Thwaites wheels home in a large suitcase that falls apart en route.

Guided by historical descriptions of early smelting, over many attempts, Thwaites eventually builds a "furnace" that recovers a product resembling usable iron. But the product isn't of high enough quality and Thwaites resorts to a microwave-powered furnace to obtain a usable product. Thus, just to get to the point of having workable iron, Thwaites has resorted to many shortcuts involving the help of experts and modern-day devices; yet all these shortcuts are only the first steps in the process of making an operational toaster. The full story is far too long to even summarize, but it's not surprising that Thwaites continues to take many more shortcuts.

Thwaites summarizes his experience with the toaster by saying, "My attempt to make a toaster has shown me just how reliant we all are on everyone else in the world. Though there's romance in that idea of self-sufficiency and living off the land, there's also absurdity. There is no turning back the clock to simpler times—not without mass starvation anyway." Thwaites is an unusual individual, having later tried to live in a pasture as a goat, but there is little reason to believe others would do much better; they might well do worse without his disarming charm with strangers.[42] His final toaster survives in all its glory as an exhibit in the Victoria and Albert Museum in London (figure 4.3).

Other authors with deep backgrounds in science have tried to envision what it would take to rebuild a high-tech society after a disastrous collapse. Here, too, the need for groups of specialists embedded in functioning communities is far more intense than is immediately obvious. Lewis Dartnell, a science writer with a research background in biology, illustrates in his book

Figure 4.3
Part of Thomas Thwaites's *The Toaster Project*, on display in the Victoria and Albert Museum. *Source*: Photo by The Wub (Wikimedia Commons, CC BY-SA 4.0).

The Knowledge the challenges of rebooting modern technology after a disaster such as an asteroid strike.[43] Even if one out of a hundred people survived the disaster, after a few generations of trying to live off of what was left by those who perished, all preexisting technology and supplies would have deteriorated so badly that the remaining survivors would have to reinvent the technological world from scratch.

Because Dartnell's analysis envisions communities developing specialists with deep knowledge in specific areas, the challenges posed are less than those of one person acting on their own; and because everyone vaguely remembers what once existed and what those things did, they know what goals are achievable. Yet even with all those advantages over an individual, reinventing the modern world is immensely difficult. Dartnell takes the reader through the details of what it takes to resurrect a modern civilization. He starts with agriculture and the thousands of years it took to select high-yield grain seed. If the international seed banks were not available, people might well starve before they could make enough progress to survive.[44] Even then, the right crop must be chosen for a given location, climate, and season. Appropriate soil is also critical, as are the best ways of preparing, irrigating, and harvesting. Just the initial agriculture section of Dartnell's book quickly reveals how utterly dependent we are on expertise clusters in the present and the accomplishments of specialists from the past. The convergence of conclusions drawn by authors of disaster recovery scenarios and by the study of cumulative cultural evolution is striking.

After agriculture, challenges and interdependencies of similar enormity are revealed in storing and cooking food and making useful clothing. Finding and gathering substances such as lime, various oils, and the ingredients of soap is not much easier. Learning to use the right kinds of wood and clay also presents huge hurdles to a novice. The hurdles become higher and higher as Dartnell discusses restarting medicine, reliable power systems, transportation, and communication systems. Even larger challenges must be overcome to reach the level of technology of the present day. From the earliest stages of technology growth in hunter-gatherer communities, progress depended on the ability to accumulate and share knowledge across generations. The simplest multipart tools, such as hafted axes and bows and arrows, did not appear because a full-fledged idea suddenly emerged in the mind of an individual.

Try on your own to build a durable and effective hafted axe from raw materials in the wild over the course of a weekend. You must pick the right kinds of wood for handles, stones for axe heads, and fibrous materials for binding them together. You also must figure out appropriate sizes and shapes for each component, how to modify the initial materials into functional parts, and how to bind them together in ways that endure over many blows. Without any help from accumulated knowledge about all these parts and their interactions, and without any other tools to modify parts, the challenges are overwhelming even when you have the huge advantage of having seen working examples.

Converging evidence of human propensities to underestimate the paths and partners involved in creating technological revolutions can be found in how badly we underestimate the time it typically takes from first making a working model of a device to arriving at a version that is practical and affordable for everyday use by all members of a culture. We may assume the first light bulb and the first steam engine were quickly translated into widely available forms, when in fact it took years to progress from the first prototypes to commonplace versions. It took decades for the "invention" of commercially viable color photographs, telegraphs, telephones, AM radios, FM radios, transistors, integrated circuits, CPUs, fiber-optic cables, and LCD displays, among thousands of other examples. We grossly underestimate the amount of time needed, the number of people involved, and the reliance on historical precedents. By favoring mostly mythological stories of solitary geniuses single-handedly causing technological and scientific revolutions, we divert our attention from the much more ubiquitous and powerful processes of cumulative cultural evolution.

Reading across all these books, and many others of the survivalist genre, brings into sharp focus our extraordinary reliance on others, often through long chains of knowledge transmission. As yet more evidence for our having too little intellectual humility we can now include the tendency to underestimate the contributions of those who came before us.

* * *

From prehistory on, widespread and relatively abrupt changes in artifacts suggest ensuing changes in thought about those artifacts and about the world more generally. The shift from single-component tools to complex tools, such as the hafted axe, spawned an explosion of other complex tools in a relatively short period of time that may have been unrelated to any changes in the human brain. As humans live longer and healthier lives, grandparents and even great-grandparents are more involved in the preservation and transmission of culture to younger generations. Moreover, the increasing ability to communicate with anyone anywhere on the planet facilitates cultural transmission between experts and novices throughout the world.

The introduction of clocks and their widespread adoption in households by around 1600 helped spark the idea that the entire universe consisted of veiled clockworks. This idea was particularly notable in extrapolations of mechanical workings to living things. But construing the biological as the mechanical can distort understandings as well.

An oft-overlooked revolution occurred with the sudden rise of interchangeable parts in the nineteenth century. Interchangeable parts enabled people to think about functions and component interactions in new ways, but they may also have limited understandings of how components revealed tools as adaptations for individual users embedded in specific environmental niches.

The introduction of electronic devices and circuits in the late 1800s and early 1900s was relatively easy to understand through metaphors arising from hydraulics and simple push-pull relations. In addition, those early electronic components, such as vacuum tubes, capacitors, and resistors, could be seen, touched, isolated, and removed from a device. This in turn allowed tinkering and repair. With the rise of transistors and then, more rapidly, with the rise of integrated circuits, this mechanistic transparency vanished. Soon the number of parts in some chips were in the millions, then in the billions. Not only did this increase overwhelm human reasoning about components and their interactions; it also led to a transition from thinking about mechanical interactions to focusing on digital operations and functional sequences.

If the intelligence of individuals reflects cultural intelligence handed down and culminating over successive generations in a distributed manner and not in the mind of any one individual, today's children may be at risk. Recent technological changes have happened so quickly that parents today are cognitively disconnected from modern digital devices; they may be ill-equipped to transmit the new forms of information needed by the next generation.

I have suggested that children today may be increasingly confined to considering the surfaces of things and their uses or functions. Information that was easily available a few decades ago has largely evaporated without other new forms of deeper information becoming available. Younger children might be especially deprived of information as their reading skills limit their ability to look for answers online. It would be remarkable if such dramatic changes in exposure to the underlying architectures of devices and in talk about them had no downstream influences on explanatory understandings and how we think about artifacts. We now need to understand whether such influences are likely to be negative, positive, or largely inconsequential.

Well before they begin schooling, children are driven to look beyond the obvious. In the past they have explored and uncovered hidden parts that provided pleasing insights into the underlying causal machinery governing the appearances and behaviors of things. As children continued to explore causal structures in the world, they came to appreciate ever more articulated explanatory signatures. This is where some of the most striking developmental growth can occur and where guidance by knowledgeable adults may be most powerful. It may also be where changes in the content of children's experiences and in the language around them can shift their explanatory preferences and practices and put them on intellectual paths far different from those of prior generations. To help us examine this possibility, chapter 5 turns to the explanation-seeking drives found in all humans from infancy onward.

5 What's Within and Why?

When our children were young, we took them every year to the open house at Cornell's College of Veterinary Medicine. While many exhibits were interesting and fun, for our family and many others there was one showstopper: the fistulated cow. Fistulated cows demand your attention. At first it's not clear what all the fuss is about. A large cow stands tethered to a post, munching on hay while looking calmly at the visitors. But then the hole comes into view. In the middle of the cow's side is an eight-inch-wide round opening, surrounded by a smooth rubber rim, that can be sealed with a lid. This alone is startling, but the real surprise is that there seems to be nothing inside. The hole is over the cow's rumen, the largest chamber of a cow's four-chambered stomach. When he first saw the cow at age three, our son Derek was completely fascinated by this odd mix of animal and machine. His interest only grew when the guide held a flashlight to the hole and invited Derek to put his hand inside.[1] Derek then looked at us and shouted that it was empty, then asked how it could be empty. As we dragged him off to the next exhibit, we tried to explain. We said that cows had several stomachs and that the biggest part, with the hole, was like a pot in which the cow made juices that it used to break down food and make it easier to digest. We assured Derek that most of the cow was not empty, and that it needed that one part to be empty to work with the other parts. This little event, reenacted with our other two sons, Dylan and Marty, captures much about young children's interests and explorations. Their intense interest in hidden workings has been a well-documented and surprising outcome of research over the past few decades.

To examine how changes in technology can influence young children's minds, we must first understand what young children want to know and what they can know. If certain kinds of things are beyond their cognitive grasp, and if recent technological changes are beyond their early cognitive capacities, those changes might influence them only modestly, through the actions of older people. Certainly, some changes in the world can influence adults' lives but go largely unnoticed by young children. Fifty years ago, most people did their income taxes by hand or hired an accountant, while today more than 50 percent use some form of tax preparation software. The software makes the tax filing process much easier, but it has little impact on children. They have neither the mathematical nor the cultural knowledge needed to complete a tax return.

Does much of the world similarly exceed young children's cognitive capacities? Conceivably the decline in explainable insides is irrelevant to young children both because of its cognitive complexity and because it doesn't concern them. If the mechanical patterns of causation in most everyday devices fifty years ago were beyond the conceptual and motivational reaches of young children, it hardly matters if those patterns are now no longer available to children for observation, inspection, and manipulation. Early theories of cognitive development typically described children as proceeding through stages involving sweeping shifts in cognitive abilities. Young children were depicted as incapable of higher-level thought.[2] They labored under cognitive deficits that rendered them unable to grasp the relational patterns found in causal explanations. If younger children were so limited in their abilities, the disappearance of insides affording mechanical explanations for adults would have little meaning for them. The evidence, however, suggests otherwise.

Researchers now describe children as deploying a diverse array of cognitive skills to rapidly learn about everything from dogs to doggerel. In thousands of studies conducted around the world, young children and preverbal infants have shown sophisticated cognitive abilities even as they also show limitations compared to older children. Several of these abilities reveal early emerging quests for explanations. These attempts to uncover causal patterns suggest that an environment devoid of revealed mechanisms could well have both immediate and long-term cognitive consequences. This chapter examines children's interest in hidden or nonobvious causes that explain what they encounter. Chapter 6 considers early thought about the worlds in which objects are embedded, most notably their social world.

Children's minds are pivotal to understanding how knowledge grows and supports explanatory ability on a lifelong basis. Young children have cognitive biases that channel them toward certain forms of interpretation and ways of explaining the world. They provide revealing clues about how to make useful causal interpretations with impoverished knowledge. That ability in turn informs accounts of how information is transferred across generations.[3] By focusing on how children start to uncover the intricate causal webs that enable them to build explanatory understandings, we learn how children go beyond the surface of things to grasp underlying causal patterns. We see how notions of causal complexity emerge and why they matter. We see why a particular form of causal cognition—mechanistic thought—may have a special role in children's thinking early on. These developmental stories show different patterns across broad domains of thought, such as naive physics, biology, and psychology. In turn, those differences bear on how children's emerging explanatory insights might be influenced by generational shifts in their ambient technologies.

Early Drives to Uncover Explanations

What drives infants and young children to look inside, lift off covers, and manipulate what they find? Answering this question helps us know whether such actions are becoming a useless, outdated practice from earlier eras of human history or remain important today. If it is increasingly common for children to look for more information and find that the uncovered parts provide no insights, we need to know the immediate and long-term effects of failing to achieve explanatory closure.

The drive to find explanations has been described by the psychologist Tania Lombrozo and colleagues as *explanation-seeking curiosity* (ESC).[4] ESC goes beyond the notion of curiosity as the desire to acquire more information of any kind, whether that is facts, experiences, or visual displays. In seeking explanations we expect a certain type of information, usually of the sort that answers why and how questions. In addition, we often want explanatory information to provide feelings of learning and utility. All these cognitive activities are present in young preschoolers.[5]

Because the search for explanations often transcends mere curiosity, it is better understood as an act of wondering. Wondering here includes not just wanting an explanation but engaging in conjectures about what that explanation might be.[6] In wondering about how a trail camera turns on when wildlife walks by, we might wonder whether the triggering event is a sound, an image, or an infrared profile. We rarely want an explanation without having some expectations of what plausible explanations might look like. The more we know, the richer the array of possible explanations we might have in mind, either explicitly or implicitly. We can be surprised by an explanation that is far different from anything we expected, but that surprise can help guide future explorations.

Insights in Infancy

Studies over the last two decades consistently characterize young children and infants as sensing that insides are important. Young humans also have a specific interest in orderly clusters of parts that reveal mechanisms. To grasp mechanism, however, they must also distinguish causation from correlation. Causal relations are the glue that hold explanations together, and they are detected early in development.

The Causal Connection

In the first few months of life, infants preferentially attend to causal patterns. One object colliding with another is more interesting than two independently moving objects. Moreover, observing a recurring interaction between two objects can trigger a search for underlying causes.[7] Causal mechanical interactions are especially salient amid the many kinds of physical interactions infants might observe. Researchers still debate whether the first ways of grasping such interactions are based on perceptual systems that ultimately result in more conceptual notions of causality or whether the idea of cause is part of our conceptual architecture right from the start.[8]

Three-month-old infants can construe in causal terms actions that are beyond their own abilities. If they watch a person act on something in a way that they couldn't possibly do themselves, such as moving a large object from a high shelf, they nonetheless see the action as goal-directed.[9] Infants at least as young as eight months will infer causation from contingencies even when the causal interaction is novel and unusual.[10] Across multiple studies, infants have been found to detect causality of the sort found in classic physical mechanics as well in social interactions of goal-directed agents.[11] Several research groups are asking how these different senses are interrelated in early development.[12] A central issue is whether one sense of causality is prior to and a foundation for the others or whether several specific senses develop independently early on and only merge later to be understood in the most general sense of "cause."

The Insides Story in Infancy

Infants think early on that inside parts are special as compared to surface parts. This bias has often been described in older children and adults as a version of essentialism, the belief that a hidden inner property defines a category.[13] Based on prior work from our research group showing that older children believe in essences in animals, such as not being fooled by superficial changes that made one animal look like another, we became interested in earlier origins of such beliefs.[14] When George Newman was in our lab, he led a series of studies showing fourteen-month-olds video cartoons of cats "dancing" in different ways.[15] One cat had a blue hat and a blue belly while the other cat had a red hat and a red belly (figure 5.1). (By making the cats' skin translucent, the belly appeared to be inside and the hat outside.)

The red cat danced by bouncing up and down and the blue cat danced by swaying from side to side. These movements were repeated several times in presentations to infants. Then the critical test trials were presented. Infants looked longer at cases when the cat with a red hat and blue belly

Figure. 5.1
Dancing cats with colored hats and bellies. By at least fourteen months of age, infants expect particular dances to be the same in cats that have bellies of the same color and hats of different colors, but do not expect them to be the same when the hat color is the same but bellies are different. *Source*: Newman, G. E., Herrmann, P., Wynn, K., & Keil, F. C. (2008). Biases towards internal features in infants' reasoning about objects. *Cognition, 107*, 423.

danced the original red-red cat dance than when the cat with the blue hat and red belly did. In most infant studies of this sort, looking longer is seen as a reaction to an unexpected outcome. Older humans call this being surprised. More broadly, the infants more strongly expected cats with the same insides and different hats to engage in the same dance style than the cats with the same hats and different bellies.

Intuitions about insides were then probed in another way to ensure that infants weren't merely responding to the relative ease of removing hats. Infants at fourteen months were introduced to a translucent physical object that resembled an animal with eyes and ears. Inside the object was a white box. Thick green stripes resembling hair were painted on top of the object's head. In this case, the properties of the object both internally and externally were permanently connected to the object. They were not normally removable like a hat is. In one condition, the object moved on its own and made a noise. In the second condition, the object was moved by the experimenter and shaken to produce the same noise.

After witnessing these two events, the infants were given a choice between two new objects: one object had the same internal features but did not have the external paint pattern while the other object had the same external paint pattern but no internal features. Which object would the infants choose? If infants assumed insides were special, those who saw the self-moving object should pick the object with the same interior, assuming that those insides were important to understanding how it moved and made noises. In contrast, infants who saw the manipulated object should not show a preference for the insides because there was no reason to see the insides as responsible for either motion or sound. Indeed, the infants showed those preferences.

Other research groups have also shown linkages between insides and "behaviors," broadly construed. Across several studies, the effects seem to be strongest when the objects involved had some features associated with animate creatures, such as eyes or fur.[16] Animal-like features may have suggested goal-directed animals to the infants. But the idea of "animal" early on is more skeletal than that held by older children and adults.

Goal-directed agents are not the same as belief-bearing ones, and the difference is important in understanding how infants think about the insides of self-propelled things. Consider how an adult human might think differently about a bacteria swimming toward a nutrient and Romeo drinking poison because he holds a false belief that Juliet is dead. Many bacteria have chemoreceptors that detect a nutrient gradient. These receptors in turn activate signals that make flagellar motors turn clockwise or counterclockwise. launching a series of movements that ultimately drive the bacteria toward the stronger concentrations of the nutrient.[17] Romeo's action was also initiated by detecting information through sensory receptors and was mediated by a series of chemical signals, but such factors seem largely uninteresting in comparison to how beliefs explain his actions. Do infants understand the two kinds of agents as different? In some tasks, they do respond in ways suggestive of an awareness of false beliefs; however, the strength and generality of that awareness remains unclear.[18] We don't yet know if they can sense some agents acting merely through a chain of physical events and others acting on beliefs and desires.

Thinking about insides as causally responsible for observed properties, such as motion or sound, requires a sense of containment. This involves knowing how objects can be inside other objects and how objects can be revealed by lifting a cover. Infants know this by at least three months of age.[19] An understanding of containment may be present early in development because it is a directly perceived relation in situations where a lid is partially open and the object can be seen inside through the gap.[20] Containment also may be first understood as part of the set of early expectations about self-propelled social agents. Such entities can simultaneously be multiple kinds of things, such as physical objects, goal-directed agents, and belief-bearing agents. Infants may not be explicitly aware of these multiple class memberships, but they may nonetheless benefit from the profusion of cues that arise from one thing being simultaneously a member of several broad domains.

The psychologist Rene Baillargeon and colleagues have proposed the following integrated account of what infants know with respect to insides and broad classes of kinds.[21] By twelve months, infants distinguish self-propulsion from other forms of motion. They know that a self-propelled agent has a form of internal physical energy that enables it to produce physical forces or resist them. Adults might think of the critical insides as a spring, but infants probably have little more than abstract expectations about something producing or resisting forces. In contrast, agentive objects have mental properties, such as goals, that motivate and guide its actions. When objects are understood as both self-propelled and agentive (as indicated by, say, furriness and goal pursuit), Baillargeon argues that infants think of them as animals and have strong expectations about important inside components. Three causal reasoning systems are assumed to be present in the infant's mind: a physical reasoning system with the notion of physical energy, a psychological reasoning system with the notion of mental states, and a biological reasoning system with the notion of innards. If this account is correct, it raises questions as to how children think about the inside parts of nonliving things, such as mechanical toys whose inside parts cause distinct behaviors.

Variations on this theme of infants placing a special importance on insides and hidden causes have repeatedly been found.[22] Moreover, this early bias for insides is surprisingly abstract, as not only infants but also young children know almost nothing about the specific details of those insides.[23] Infants as young as six months expected that self-propelled objects had insides that mattered to the objects' motions. When they observed an experimenter open the object and inspect and manipulate the insides before closing it again, they expected the "operation" to modify its future movements more than after watching the experimenter simply pat the object on the outside.[24] They had no specific expectations about the insides, just that they were causally important to motion. In line with previous work, the researchers also suggested that the infants assumed the self-moving box was a biological entity.[25]

As they grow, infants engage in more active exploration and manipulation of objects. They make things happen and observe the consequences. They tinker with things they encounter, actively exploring different options. The psychologist Lisa Feigenson and colleagues had eleven-month-old infants observe an event that seemed to violate physical principles of solidity or gravity. The infants didn't just notice the violation; they explored the offending objects more carefully, seeking an explanation for it. If the object appeared to pass through a solid wall, they pounded it against another hard object to see if it really was solid. But if it seemed to float in air and defy gravity, they repeatedly dropped it to see if sometimes it didn't fall. Finally, immediately after noticing the violation, infants were also taught a new property of the suspicious object: a novel noise. They learned the noise better when it was made by an object that violated basic physics than when it wasn't.[26]

Feigenson's group also studied eleven-month-olds' reactions if an explanation was provided for a violation. A wall that looked solid from the side was turned to reveal a tunnel in the middle through which an object could fit. As soon as the infants saw this "explanation," they stopped exploring the puzzling object.[27] These infants were not just curious and eager to see something else; they were also seeking a specific kind of explanation, which, when made available, immediately satisfied them and caused them to stop exploring. To distinguish this behavior from mere curiosity, we can again think of it as a form of wonder.[28] While this study used explanations that were visual instead of verbal, they certainly appear to explain a previously puzzling phenomenon.

Children as young as two, barely past what we think of as infancy, show signs of encoding situations in explanation-friendly formats. When shown events in which a gumball machine delivers a ball whose color was either highly unlikely, given the rarity of that color, or impossible, given clear evidence that no such color was present, they appeared to encode the impossible events differently from the highly improbable ones. Thus, they showed greater learning after impossible events.[29] Enhanced learning is thought to occur because impossible events forced them to realize that their model of the situation was simply wrong and had to be changed. They seemed to be engaging in logical reasoning about the situation rather than simply adjusting probabilities. This interpretation fits with a series of recent studies by my colleague Nicolo Cesana-Arlotti that strongly suggest the presence of deductive reasoning in infants and young toddlers.[30] For example, in one set of studies, preverbal twelve-month-olds were first shown two objects with identical tops but different bottoms, such as a dinosaur and a flower where the top halves showed a sequence of bumps that was either the scales on a dinosaur's back or the petals of a flower. Then both bottom parts were covered such that the infant just saw two identical bump sequences over the barrier. The experimenter then revealed one of the objects, for example the dinosaur, and, after a pause, revealed the second object. When the second object was another dinosaur instead of the flower, the infants looked longer than when the flower was revealed; they seemed to be going through a logical reasoning sequence that ruled out the possibility of a second dinosaur. They seemed to be thinking something like "The hidden object is either the dinosaur or the flower; the dinosaur is ruled out; therefore, it must be the flower."

Children may leave infancy with a more articulated set of expectations about the properties of biological agents than of artifacts. But this changes rapidly as they learn about the mechanical underpinnings of various devices. Direct experience with the working parts of artifacts may be central to early growth of knowledge about what makes artifacts do what they do.

Thinking About Insides After Infancy

After infancy, children continue to build on their understandings of insides. They typically start at abstract levels with little knowledge of particulars. They have expectations about internal stuff that causes actions or produces sounds but know little about the perceptual details. This early emphasis on insides leads to new views about the nature of essences. George Newman led a series of studies showing a transition during the elementary school years from thinking about essences as a single focal point in the middle of an animal to thinking about essences as distributed throughout the animal.[31] Children were asked what inside parts of an animal would help scientists know what it really was. Children younger than six years favored just one spot in the entity, typically near the center of its torso. Over the next few years, children gradually came to believe that parts from almost any location inside the animal would work. When we looked across a wide range of naturally occurring things, children shifted first for minerals and elements (e.g., gold) and two to three years later for animals.

The shift from focal to distributed essences for animals is impressive considering the real causal pattern children are trying to interpret. Small samples taken from almost anywhere in an animal's insides can identify the kind of animal, because almost all the samples would contain sequences of DNA that are uniquely linked to an individual species. To fully understand why every part of an animal can be used to identify the species requires some sense of how animals grow from a single fertilized egg to become multicellular organisms with each cell containing the same sequence of DNA. The DNA does not directly predict the appearance of each cell the way that tiny portions of gold predict the appearance of a lump of gold. Instead, understanding the relationship between a tiger's overall appearance, behavior, and DNA requires knowing how tiny cells compose full-grown tigers and how the initial fertilized egg subsequently produces cells containing the same genetic "recipe." Clearly, schoolchildren are not shifting because they acquire this level of understanding.

We don't yet know what drives children in the middle of elementary school to change where they locate essences in animals. They might be influenced by language patterns that imply a distributed essence—a concept that is difficult to grasp but becomes more compelling over time. One suggestive piece of evidence for a shift mediated by language comes from a study of how large language models in AI systems such as ChatGPT predict category membership when given a description in which an animal is transformed into one that looks different. The program predicts the category using an apparent notion of essence.[32] Since these models are making predictions based on massive datasets of language samples, it seems that the notion of essence is latent in and inferable from language use patterns. Such patterns may not cause an essentialist bias, but they apparently carry information implying essences. They may also carry information implying that bits of essence are distributed throughout the entire entity.

Mechanical Parts

Toddlers reveal expanding understandings as they start to put their wondering into words, asking why and how questions and often speculating on what makes a thing work or move. These questions and explorations may be triggered by witnessing spatially sensible clockworks resembling those described by Archimedes—things like levers, pulleys, springs, and wheels. They ask more questions of this sort about artifacts than about organisms.

Our three sons and four granddaughters have all been enthusiastic explorers as infants and toddlers. As soon as they could walk reliably, they would rush up to novel things with outstretched hands and tweak, open, or tap objects of interest. As spectators, we felt the highlights happened when those actions resulted in an apparent "reveal." Their eyes opened wide; they usually smiled and repeated the effective behavior many times. These little explorers displayed apparent insight as they discovered noise-making beans inside a maraca or uncovered roots that anchored a flower in a pot and proudly presented the flower with dangling roots to others.

By four years old, children know that novel objects with different internal parts are likely to have different causal properties. If those internal parts are transferred to another object, four-year-olds infer that causal properties transfer as well. That is, external causal properties stay with the original insides.[33] This pattern is found in children roughly three years older than the fourteen-month-olds who match cat insides to cat behaviors, and it is different in scope. Earlier reasoning about inside parts and external properties tends to be strong and reliable when done with entities that are seen to be agentive animals. The four-year-olds in the study also noted correlations between external properties and part particulars of novel mechanical artifacts. Thus they might note a correlation between a blue triangular block inside an unfamiliar box and a specific sound emitted from the box. They assumed that changing the blue block was more likely to change the sound than changing the exterior of the box. This ability does not suddenly come online at four years, and simpler versions may occur in late infancy or among young toddlers. As we learn more about how the ability unfolds and grows from age one to age four, we will get a better sense of what kinds of experiences can accelerate its growth, such as certain kinds of interactions with mechanical devices of just the right level of complexity.

While infants may start thinking first about objects and their insides with a vague notion of agentive animals, when they start doing the same for artifacts, as toddlers, they have the potential to learn a great deal more about mechanical artifacts through exploration both as individuals and in the company of peers or more knowledgeable adults. Their knowledge of the relations between animals' innards and outsides, in contrast, may plateau as it becomes more difficult to learn more either through direct experience and exploration or through instruction by suitably knowledgeable adults. Tinkering with the insides of a mouse to see how it moves is less feasible than tinkering with the insides of a windup car to understand its motions. Mechanical knowledge and understanding may therefore provide a critical set of models for conceptual growth in early childhood and beyond; those models may then cycle back to inform more advanced reasoning about the insides of living things.

Inferring Insides from Outsides: The Case of Complexity

Devices vary greatly in mechanical complexity. A pair of pliers is much less complex than a grandfather clock. The obvious differences suggest that estimating mechanical complexity is a trivial process. Perhaps young infants sense mechanical complexity by estimating the total number of moving parts. But that strategy will not work with real artifacts. A beanbag chair may have many thousands of moving parts (beans or modern substitutes), but it is mechanically trivial in comparison to a push lawn mower, which may have only a few dozen moving parts. Mechanical complexity requires a sense of the structured causal interactions between different parts and can be difficult to define in terms of a set of rules that an automated program could implement. Mechanical complexity is also distinct from visual complexity. A simple one-piece shield may have a dizzyingly complex painting on its surface but be mechanically trivial. As shown in figure 5.2, an intuitively mechanically complex thing is more than the number of parts, number of distinct parts, or diversity of functions of parts. All of these must be integrated into a causally coherent sequence.

Having a sense of the mechanical complexity of devices could be useful to laypeople. They will make mistakes of both overestimation and underestimation, but being able to form roughly accurate assumptions could serve them well. If you know that a broken device is mechanically simple, you might try to fix it yourself, or you might use your knowledge of what is required to fix it when getting repair estimates. If you know that a device is extremely complex, you may devote your energies instead to finding the most relevant experts and knowing how to rule out unqualified individuals. When evaluating the cost of a device, knowledge of the cost of its raw materials and its mechanical complexity may provide a sense of a fair price for the assembled device.

Do people agree in their intuitions about mechanical complexity? And if they do, at what age do they first show agreement? Jonathan Kominsky led a project in our lab that addressed this question.[34] People need not know a mechanism in detail to sense its complexity. Adults believe jet engines are

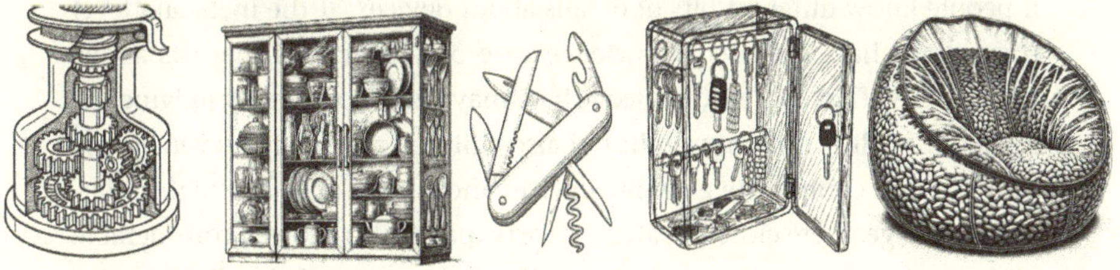

Figure 5.2
Mechanical complexity is apparent in ways that do not simply count the number of parts, assess the visual or functional diversity of parts, or tally connections that link parts together. The sailboat winch on the left is obviously more mechanically complex than the other four objects even though we have no precise rule for determining mechanical complexity. *Source*: author.

much more complex than eggbeaters not simply because jet engines have many more parts, but also because they know that jet engines must have several subsystems that serve different functions, each of which is likely be mechanically complex. They might know that jet engines need a fuel supply system, spinning turbines, and combustion zones, and that each system has function-specific parts. Adults have incomplete knowledge—most probably don't know about jet engine compressors—but they have useful shared senses of the complexity of most devices and machines.

Because the knowledge guiding the judgment of mechanistic complexity is sparse and fragmentary, we called the information represented *mechanism metadata* by analogy to the familiar concept of web page metadata. Metadata simply means data about data. For a web page it includes information that will be useful for others who might want to access that web page, such as the size of the data, when and how it was created and for what purpose, the creator of the data, and the quality of the data. For a photo, metadata includes information about the size of the image, its resolution, and the shutter speed used to capture it. Text metadata includes the amount of text and a summary of the content. All these types of metadata help search engines organize and sift through web pages.[35]

For mechanisms, we characterized metadata as including information about the broad category (e.g., artifact, organism, nonliving natural object, substance), the approximate number of discrete parts with different functions, the degree of diversity of functions, and important causal patterns (such as that an engine is often the common cause of many effects). Mechanistic metadata also helps guide searches for relevant human experts. Even if people know different bits of details about devices, at the metadata level they might have a shared knowledge base of abstractions that sits above such details. That shared knowledge base may guide complexity judgments.

In our studies, not only adults but also children as young as seven agreed on what was complex (e.g., X-ray device) and simple (e.g., flashlight). Ratings at all ages were closely matched. Participants also showed considerable agreement within each age group on when help was needed to fix an entity when it was defective and the amount of help needed (while also showing some modest age differences in how closely help was related to the two complexity ratings). Finally, when ratings of complexity extended to body parts (e.g. teeth vs. eyes), consensus was again found; the consensus was not as strong, but it was still above chance levels. This extended complexity intuitions beyond mechanical artifacts. While consensus was present at the younger ages, it became stronger as children acquired more shared information about the world. Complexity intuitions also become more intricately connected to expert seeking and evaluation. These changes are presumably mediated by ever richer knowledge of how inside parts and systems are causally connected to observable properties.

Behavior and Complexity

How do children understand the relations between the surface complexity of actions and behaviors and internal parts? The psychologist David Sobel and colleagues showed three- and four-year-old children two novel devices. A button push turned on both, but one device showed more variable effects. The simple device turned on a steady light of one color. The complex device's light cycled through a sequence of several different colors.[36] The children were then asked to match two pictures of possible insides to the two devices. One picture showed insides with fewer features and fewer types of features than the other. Four-year-olds matched the more complex interior to the more complex effects, but three-year-olds did no better than chance even in later experiments that attempted to make the task simpler. The researchers speculated that, between three and four years of age, mechanistic knowledge becomes integrated across artifacts and general patterns relating to complexity become more apparent. By age four children's representation of causal structures may be seen as "mechanism laden," such that mechanism may come to dominate much of their reasoning about artifacts from that point forward.

Complexity in these studies was truly abstract, as the interiors could be interpreted only in terms of the number and variability of parts. No actual mechanical interactions could be envisioned; all the parts were electronic components whose functions would be opaque to all adults except a few highly specialized experts. Ultimately, understandings of mechanisms should also include notions of plausible versus implausible mechanisms—intuitions long known by researchers to be held even by preschoolers.[37]

Preschoolers therefore appreciate that internal complexity of parts is an indication of behavioral complexity. This seemingly simple idea is surprisingly intricate when considered fully both in terms of what aspects of behaviors guide complexity inferences and what abstract properties of interiors are seen as indicating complexity. In terms of behaviors, all of the following can drive complexity: number of discrete behavioral types; diversity of behaviors; and the precision, variability, and timing of behaviors. In terms of interior properties, indicators include number, heterogeneity, relational structure, and stationary versus moving parts.

Consider the case of diversity of behaviors. Suppose you encounter two machines: One makes cupcakes and muffins; the other makes cupcakes and soup. Not knowing anything else, is one machine likely to be more complicated than the other? Rick Ahl led a series of studies in our lab asking how these intuitions develop.[38] Adults judged the machines that process more diverse things to be more complex because they assumed more machinery and more parts of various kinds were needed to process more different things. By age six, children also used diversity of products or behaviors to infer more mechanistic complexity. Four- and five-year-olds did not make a strong link between the diversity of external behaviors or functions and internal complexity. However, based on informal interviews with other young children, children as young as four might grasp the diversity–complexity link if the stimuli contrasts are more compelling.

When children succeed in these tasks, they do so with minimal knowledge of the details of muffin or soup makers. Yet they sense that a machine making cupcakes and soup will require more internal equipment than one making cupcakes and muffins. When the children in our study succeeded, they were equally successful across several machines processing different diverse and similar things (e.g., one that cleans sweaters and pullovers versus one that cleans sweaters and glasses, or that picks daffodils and daises versus daffodils and raspberries). To succeed in this task, children must have subtle senses of the natures of the objects, the actions performed, and the intricate interactions between the two. While a large flat brick might be an excellent device for stopping doors from swinging, holding down stacks of papers, and protecting surfaces from hot pans, these diverse uses do not imply mechanical complexity. By age six children grasp the difference.

In follow-up studies, Ahl explored what properties of inside parts were the strongest drivers of young children's intuitions about complexity. He first showed that even four-year-olds' intuitions about relative complexity were aligned with older children's and adults' intuitions. Ahl then contrasted three internal cues to internal complexity: number of components, diversity of parts, and density of connections between internal parts. For example, simple insides might comprise one battery and one small cylindrical part while complex insides might have a battery with five differently shaped parts and several connecting wires between them. Number was tested while keeping diversity and connectivity between parts the same. Similarly, diversity and connectivity were tested while keeping the other two cues constant. In these studies, comparing the three cues to complexity, developmental change was revealed. Before age six, children used only number to estimate complexity. When six-year-olds did start to use cues beyond number, they used connectivity and diversity at roughly the same time. Once they started to notice cues beyond number, they seemed to be more actively searching for other possible sources of information.

In a final set of studies, Ahl asked children what sorts of actions on objects would be helpful in assessing the objects' complexity and what sorts of actions would not be helpful. At around seven years, children judged actions such as difficulty fixing an object as more useful than actions such as difficulty in spelling an object's name. Children six years and younger struggled to tell which of these actions was more informative. When the uninformative actions were made even more irrelevant in a follow-up study, children knew those extreme cases would be less informative about complexity. These included such actions as seeing the device at night, putting a piece of tape on it, and singing a song close to it. In this second set of cases, nothing about the action could be connected to anything specific about the object, and at least by age five children knew that such actions were of no help in assessing complexity.

Taken together, the intricacy of these intuitions about internal complexity is impressive. Without any explicit instruction about how to assess complexity and with no experience with a given device, young children already have some sense of how to infer complexity from abstract relations between unseen internal parts and surface properties. These intuitions have the potential to grow dramatically in the early school years, but only if children's environments contain devices with interpretable interiors.

Clockwork Cognition: Early Interest in Mechanistic Explanations

Well before they start elementary school, children have a particular affinity for mechanistic explanations. Young children see causal mechanistic information as providing a privileged basis for inferences about the quality of other people's knowledge.[39] Several studies led by Kristi Lockhart in our lab showed that young children understand the special value of mechanistic information as opposed to functional information or facts.[40] For example, in one study children heard about why two people favor different brands of electric cars. One person favors brand X because its engine helps it speed up extra quickly, its special springs reduce bouncing on bumpy roads, and it has a powerful battery that lasts a long time. The other person favors car Y because it comes in great colors and is shiny; its seats are soft leather that is very comfortable, and it has lots of cupholders for drinks. When asked which person knew more about cars, even five-year-olds chose the person who favored car X. They sensed that the mechanistic information entailed having a more complex body of knowledge.

More research is needed to fully characterize how preschoolers and young schoolchildren think about causal mechanisms and why they value this form of information. They attend to mechanistic information not only for artifacts but also for living things by at least age four.[41] When shown two different explanations as potential answers to a question, they prefer mechanistic explanations over fact-based ones. Thus, in judging explanations for how fish breathe under water, young children viewed mechanistic explanations as more helpful than descriptions of properties such as how long fish could stay underwater. These children extended a common way of explaining properties of artifacts to organisms. This extension might depend on receiving sufficiently rich experiences with artifacts whose internal causal mechanisms can be deciphered through visual inspection and tinkering.

Researchers in science education have long stressed the importance of mechanical reasoning.[42] Spatial cognition figures prominently in many of those discussions. When thinking about components of devices, we often visualize causal interactions between objects over space and time. Moreover, having people practice with problems in spatial tasks seems to enhance mechanistic thought.[43] Practice in spatial cognition also enhances mathematical learning.[44] Children as young as three years old show signs of such benefits, with strong effects repeatedly shown by six years of age. Even simple exposure to spatial language promotes spatial thought in four-year-olds. Using more words referring to shapes (e.g., *square* and *circle*) and spatial relations (e.g., *over* and *under*) was sufficient to enhance success in tasks such as mentally rotating objects.[45] From a neuroscientific perspective, the same neural circuits that initially evolved for spatial navigation have been argued to "provide the building blocks for many core elements of human thought."[46] Humans may have grafted mechanical reasoning skills on top of spatial ones, leveraging their efficient and powerful representational formats. Not all mechanisms are equally appealing, however. As adults we prefer mechanisms that are elegant, efficient, and well organized. These preferences have elements that appear early in development as well.

Order and Efficiency

In 1959 the chemist and novelist C. P. Snow delivered a provocative lecture at the University of Cambridge with the title *The Two Cultures*. Snow argued that universities had caused disciplines to become members of two distinct cultures: the humanities and the sciences. As a result, people in each community had no shared intellectual currency with colleagues from the other culture.[47] In support of his claims, Snow observed that most humanities scholars were unable to describe one of the most foundational ideas in the natural sciences, namely the second law of thermodynamics. The law was first formally characterized by William Thomson (a.k.a. Lord Kelvin) and separately by Rudolf Clausius in the 1850s.[48] Simply put, the entropy of the universe is always increasing. The law is often described as the idea that the physical universe is relentlessly declining from ordered patterns and relations to more random ones. A situation of increasing order can occur only as a local exception and is invariably compensated elsewhere by a greater amount of decreasing order. These principles relating to entropy and order can set an undergraduate's head spinning on first encounter. Yet preverbal infants have strong intuitions about related phenomena that involve increasing and decreasing randomness.

In studies led by George Newman in our lab, infants as young as eleven months expected only goal-directed agents to be capable of creating order out of disorder.[49] Infants watched a barrier come down in front of a neatly stacked array of blocks. They then saw a ball appear next to the barrier. In one condition, the ball had eyes and seemed to look at and interact with the viewer in a manner that strongly suggested it was a goal-directed agent. In the other condition, the ball had similar surface markings but did not act in a way that implied any sense of intention or agency. The ball then went behind the barrier for a short interval. Then the barrier was removed, revealing the blocks in a disordered array. The infants reacted the same whether the ball was agent-like or inanimate. They thought it equally reasonable for the ball and the agent to convert neat, ordered arrangements to more disordered ones. However, when the setup was reversed, such that the blocks started out as a disordered array and then became ordered, the infants looked longer at the inanimate ball; it was unexpected for an inanimate ball to neatly order a random scattering of blocks.

In C. P. Snow's defense, we did not directly ask infants about the second law of thermodynamics. More importantly, while older children and adults virtually always saw it as impossible for the ball to create order out of disorder, they were surprisingly tongue-tied in their attempts to explain why. Older preschoolers might just repeat their reaction more forcibly with responses like "Because it just can't; balls can't clean up messes." But many adults as well had difficulty putting into words their strongly held convictions. What is it about goal-directed agents that allows them to locally violate increasing randomness or disorder even if it is not possible in the universe as a whole? Readers are encouraged to try to come up with a clear and concise response that doesn't immediately raise other questions.

This finding fits well with a longer tradition of studies showing that preverbal infants see goal-directed agents as behaving dramatically differently from inanimate objects.[50] For example, if they see both goal-directed agents and inanimate objects repeatedly move toward an object in a specific location, they react differently when the object is moved to a new location. They expect the goal-directed agent to go to the same object in the new location and expect the inanimate object to go to a new object in the old location. Finally, in a study conceptually similar to the ordering/disordering experiment, infants assumed that only goal-directed agents could pluck just red balls out of a jar containing a random mixture of blue and red balls.[51] Only goal-directed agents could decrease the randomness of a collection.

Preverbal infants' preferences for goal-directed agents as the causes of changes from disorder to order are likely to be based on abstract and widely generalizable intuitions. Consider some other ways a system could go from disorder to order that would stand out for adults. Of special interest are instances with no *perceptual* similarities to other cases of ordering but with strong *conceptual* similarities. One abstract way to construe a scene as becoming more organized involves seeing it as becoming more informationally compressible. Compression algorithms are commonly used in visual image processing, but we can also think of conceptual compression algorithms. Suppose spatial arrangements didn't change but the display went from a random mix of two colors to a neat sorting of red and green. As just described, infants can sense this as an ordering event. But now think of all the ways a random mix of equal numbers of balls of two colors could be organized. They could be swapped with each other so that all balls are in the same spatial locations with all the reds are on one side and greens on the other. Or the reds could all be in the middle and greens all around the periphery. Or they could alternate so that every red ball was next to a green one.

These and many more spatial configurations could segregate greens from red while covering the same points in space. When that spatial location constraint is lifted, enormous new sets of ordering possibilities arise. For example, red-green pairs could be touching each other with much larger spaces between them. A long row where every adjacent color was different would be obviously ordered, as would an alternating sequence of three reds and three greens. Several dozen such cases are easy to create, and with a little more thought, hundreds. The total number of ways that a random array of colored balls could become ordered tends to be far larger than first impressions suggest.

Once we move beyond color to consider variations in shape, size, texture, and other features, the ways to go from disorder to clearly detectable order explode. We don't know how many of these orderings infants would notice in comparison to adults, but it seems likely that infants would detect a large number. Clearly, people of all ages cannot detect every instance of either increasing order or increasing randomness. They miss plenty of both, but they see many others. Aspects of an abstract conceptual representation driving such intuitions may be shared with prelingual infants. We can consider it as a form of information compression because rather than stating the exact arrangement of every ball, we can more economically remember the pattern by stating it in terms of a rule. For example, one rule might be that all balls are different colors from adjacent ones. Implicit compression heuristics of this sort might make such intuitions more available to infants.

Efficiency Experts

Infants' impressive abilities to detect ordering events, goal-directed agents, and physical causality helps them expand their understandings of agents, order, and efficiency over the next few years of their development. They see further and further into the intricate web of relations that explain the actions a goal-directed agent will take; and, more indirectly, they see further into the ways an artifact can be designed.

Beyond understanding agents as creating order with objects, infants also construe agents as acting efficiently to achieve their goals. In 1995 a group led by the psychologist Georgi Gergely demonstrated that twelve-month-olds expected agents to behave rationally.[52] The infants in that study observed a self-propelled small ball interact contingently with a large ball on the other side of a wall. The small ball then jumped over the wall to be next to the large ball. However, when the wall was absent, infants expected the small ball to no longer jump but to make a novel path directly to the other object. They expected it to take the most efficient route. My colleague Julian Jara-Ettinger has been at the forefront of an expanding group of researchers exploring how young children use a naive utility calculus and notions of efficiency to interpret agents' behaviors.[53] According to this approach, humans construe other agents as acting in ways that maximize subjective utility. Agents choose to engage in behaviors that are the most efficient or where the ratio of reward to loss is the highest. Utility accounts of this sort have been the basic vocabulary of formal economic theories for many decades, but Jara-Ettinger and colleagues propose that an intuitive version of utility theory is fundamental to how we interpret the actions of others. By adopting that approach, they have been able to create a formal model and predict how humans of different ages make inferences about subjective costs and rewards based on others' actions.[54]

The word *naive* is critical here because the claim is not that humans are error free in their utility calculations. People show several strong biases in their calculations that deviate from correct results in systematic ways.[55] But in many straightforward everyday situations involving interpretations of the mundane activities all around us, the naive utility calculus seems to serve people of all ages well.

Infants assume that agents are more desirous of things that take more effort to acquire. If an agent chooses to climb a ten-step ladder to retrieve a box of unknown contents, rather than a three-step ladder to retrieve another box of the same size, adults assume that the agent attaches greater value to the first object. Even three-month-olds make similar inferences.[56]

Researchers are now exploring more elaborate reasoning about how others act and interact efficiently. For example, people choose more difficult actions for a reason. The first studies focused on physical actions and the effort expended in doing them. In one set of studies, six-month-olds expected agents to take the most efficient path when barriers were always present but varied in their properties.[57] Thus, if an agent who had hopped over a high barrier was presented with a lower barrier, infants expected the agent to adjust their actions in ways that optimized efficiency: just clearing the lower barrier rather than using the original higher jump. They were going beyond reasoning that with no barrier there is no need to hop, as in the original studies; they were reasoning more subtly about adjusting behaviors to make them optimal.

Young children and infants can be inspired by their inferences about another's behaviors. In one study led by my colleague Julia Leonard, fifteen-month-olds observed adults who either easily obtained a desired object or struggled for an extended period before finally obtaining it. After watching an adult struggle, infants are more likely to show increased persistence when they try to perform a novel challenging task.[58] They seem to be reasoning that people who expend great effort at tasks get just rewards. Effort has a high payoff potential. In short, well before they enter the classroom, children are making strategic inferences not only about what to learn but also about what others know and want with respect to knowledge. Models of rational, efficient, and optimal behaviors seem to be ever present in their minds. We'll return to this topic in chapter 6 as we ask how children learn from and with others.

Efficient Devices

How do these impressive acts of reasoning about the efficiency of agents' physical and mental activity relate to the products of our actions, namely the designed efficiency of objects? To date, almost no research has directly focused on this question. One small study with adults, however, suggests close relations between the efficiency of artifacts and the efficiency of agents.[59] In that study, adults were asked to choose which of two devices was better. Both devices performed the same function equally well, but one had more internal parts involved in supporting that function. Adults judged the device with fewer parts to be superior, presumably because they saw it as being more elegant and efficient. However, when two objects were described, one with more parts than the other, and the functions were ill-defined, adults judged the one with more parts to be better under the assumption that it could do more things. Here we see close connections to the studies described earlier on how diversity of function tends to imply more internal parts and more internal complexity.

The world of designed and constructed things is intimately connected to deep-seated ideas about optimization—not only optimization for human users, such as handles of the right size for hands, but also optimization of mechanical efficiency in the device itself.[60] Questions remain concerning how adults and children discern such factors as reliability, durability, and simplicity in devices. As adults we can all appreciate caricatures of inefficient devices such as Rube Goldberg machines, but the origins of these intuitions have not been studied. The intuitions must involve the integration of several factors, including situations where a larger number of stationary parts supplies greater strength and reliability to a device even though those parts don't engage in any work. Thus a device that vibrates excessively might have several braces that strengthen it and preserve its longevity and thereby increase its efficiency. At this point we don't know whether developmental differences reflect increasing expertise with specific classes of devices or refinement of general strategies used to evaluate all devices.

An especially subtle case involves comparisons of two devices with the same function where one has far more parts and is working better. Such instances would seem to run counter to efficiency. Yet in the evolution of navigation timepieces to solve the problem of calculating longitude, each new chronometer tended to have more and more parts.[61] In many devices, from machine tools to clocks, precision requires more parts, not fewer. This seems intuitive to many of us in today's world who have some familiarity with such devices, but it is not known how we acquire such intuitions and learn to manage the tension between efficiency and precision. This leads to questions about the nature of those intuitions in younger generations who may have less experience with artifact mechanisms.

We can also ask how ideas about optimization are extended from artifacts to organisms. Do we tend to interpret organisms as acting in optimal and efficient ways? They do in some respects, but spectacular exceptions illustrate how this is a misleading strategy to apply indiscriminately to biology. Organisms often inherit mechanisms even when they're not ideal in terms of optimization. Sometimes they become irrelevant vestiges, such as the appendix, but other times they maintain the status of functional units. Even though more efficient versions might seem obvious, the evolved units constitute local minima that are unlikely to be escaped through the process of natural selection.

For example, many bacteria with flagella move toward increasing concentrations of food and away from declining ones, but they do so in a way that would be counterintuitive if they were designed artifacts. If they are heading toward a desired sugar or protein, their flagella rotate in a counterclockwise direction and they move in a straight line. But if the concentration starts to diminish, rather than simply steer a bit to the left or right, they start rotating all their flagella in a clockwise direction, which causes them to tumble randomly. They then switch back to the counterclockwise mode and start moving again in new direction. If the concentration of food there is sufficiently high, they continue in a straight line. If not, they tumble again.[62] This odd and seemingly inefficient way of moving is all the odder when one sees the working parts of the flagellar "motor," which looks strikingly like a motor and seems to confirm the mechanical interpretation. Do younger generations have a cognitive advantage because they do not so automatically slip into the clockwork interpretation where efficiency and optimization considerations dominate? Perhaps they won't be surprised by the bacteria's run-and-tumble behavior because elegant mechanical solutions do not spring to mind as readily.

Tasks involving optimizing and maximizing utilities can quickly become immensely complex. Consider the so-called traveling salesman problem (TSP). The TSP asks: If the salesperson is given a list of towns and the distances between each pair of towns, what is the shortest trip they can take that goes to every town exactly once and has them end up where they started? Figuring out the best route skyrockets in complexity as the number of towns increases. With 3 towns there are 2 possible routes to compare. With 4 towns there are 6 routes. With 6 towns, 120 routes. With 12 towns, roughly 40 million routes. With 100 towns, the routes number more than 10^{150}. Such problems abound in any situation where you have several ordered options to choose from. Another example involves packaging. Suppose you have a large collection of objects that you can select among to ship in a particular container with certain dimensions and weight capacity. What is the most valuable combination of objects that you can pack into the container without exceeding the dimensional and weight limits? As the number and variety of objects increases, the task soon overwhelms human cognitive capacities.

These examples illustrate the pervasiveness of such problems in everyday life and the potential for people working alone and in groups to gain greater mastery for more complex cases as well as to learn heuristic shortcuts.[63] But solving such problems may not always require what we assume to be highly intelligent humanlike agents. Colonies of slime molds without brains can do quite well in solving some route optimization problems, such as designing highly efficient highway and rail networks in Britain.[64] We don't know yet what humans are doing at the cognitive level across a wide range of optimization problems. A champion field trial dog who runs ten times as fast as it swims seems to optimize, in real time, routes over water and land, adjusting its route if the water is faster than predicted because it is shallow or slower because it is deep and full of weeds. Similarly, many food-caching birds and mammals (such as squirrels) use optimization strategies to increase effective retrieval later and to obtain the most perishable items first. It may be difficult to figure out the most optimal or most efficient route when the number of possible paths becomes large, but in simpler situations, where *more* efficient doesn't mean *most* efficient, humans and other animals often quickly choose a good-enough option.

* * *

We considered young children from infancy up to their first few years in school in terms of the ways each child picks up on environmental cues for how things work. Young children are well equipped for starting along that path of learning. They are intensely interested in the insides of things; more importantly, they are eager to learn about the causal relational structure that binds together internal parts and gives rise to insight about how they work as a system. We also saw how this fascination might be related to a deep-seated essentialism.

Young children have a good sense of causal complexity and mechanical complexity, and they can use several different cues to infer complexity even when it isn't directly visible. Having obtained some grounding in these areas related to causal components, young children go on to think about order and efficiency. In addition, even infants recognize that only certain kinds of entities can cause order. They realize that, at a minimum, goal-directed agents must be involved; in many cases, the agents must be belief bearing as well. Although there is no developmental research to date on how efficiency relates to artifacts, a large body of research has recently portrayed infants as closely tracking information about the efficiency of actions. They use that information to infer the goals, mental states, motivations, and behaviors of others. They also use it to infer the quality of knowledge others may possess. A recent study in which adults easily grasped the benefits of more efficient artifacts suggests that children will appreciate those benefits as well.

Given earlier considerations about how spatial cognition could augment and support higher-level reasoning and representations, children's spatial models might also support mechanistic reasoning. This possibility gains support from cases where enriched spatial cognitive experience enhances mathematical thought.

The mental lives of infants and young children go far beyond experiencing the here and now as an individual. They also learn from other people and from the communities in which they are immersed, and by transcending the immediate present and thinking about other possible worlds and what-if scenarios. Finally, they learn by stepping back and becoming observers of their own learning. These topics are addressed in the next chapter.

6 Beyond Ourselves: Other Minds, Worlds, and Perspectives

As a five-year-old, I had a peculiar understanding of how televisions worked. When someone was reading the news live, I thought they could see their viewers. After repeatedly being told the news was true in real life, I had assumed the newscaster was having a conversation with each viewer as part of a live audience. I imagined a wall of tiny televisions that allowed newscasters to scan all their viewers' faces as they read their notes (figure 6.1). I had no idea of the numbers involved and that a few hundred small screens would hardly be sufficient. I also assumed that if you did something outrageous, the newscaster might notice and call the police. I sat up a little bit straighter and was a little bit more attentive whenever I knew it was a live news broadcast.

One day, while watching the news, my twelve-year-old sister stuck out her tongue when the newscaster said something she didn't like. I was shocked and told her she was lucky he didn't see her making a face or she might have gotten into big trouble. My sister looked at me in disbelief and then burst out laughing as she tried to explain to me how silly my idea was. It took a while for me to see why my view couldn't be right. Among other tests of my view, we staged outrageous skits in front of live broadcasts to see if the announcer noticed. (My misconception might be a little more forgivable given that in 1957 only a little more than 50 percent of US households had televisions, and almost no households had more than one.)

Developmental psychologists have known for many years that younger children have difficulty taking points of views different from their own. My live newscast misconception illustrates the challenges of thinking about the self as an external observer. I would have had to step back from the immediate situation and imagine myself watching the television news along with

Figure 6.1
One five-year-old's idea of what a news anchor saw when they read the news live. *Source*: author.

many others. This perspective shift also required seeing the situation from the newscaster's point of view to gain a better sense of audience size.

Many of our biggest insights occur when we go beyond simply experiencing the world from our point of view in the here and now. We greatly expand our understanding when we appreciate how each person has a distinct journey through life and can help us see the world in new ways. Learning effectively from different people, however, often involves sophisticated reasoning about others' goals and beliefs, their points of views, and the social and physical environments in which they are situated. To resolve my confusion about broadcast studios, I had to appreciate my sister's perspective as well as the logistics of connecting broadcasters to large numbers of viewers other than myself. We are all deeply embedded in communities of knowledge and need to acquire ever more sophisticated ways of accessing and benefiting from that knowledge. Young children tend to underestimate how much help they will need from others, as anyone who has been forbidden from helping a toddler put on a jacket or open a package knows all too well. But even when they are not well calibrated, children are developing several skills for seeking out, interacting with, and evaluating others.

The growth of these essential knowledge-foraging skills depends on the experiences young children have with the world around them. The extensive body of research on how children and adults learn from others has important implications for why changes in the technological world might matter.[1] Thus a child's growing understanding connects with their cohort's knowledge and eventually comes to supersede that of the older generation in terms of what is passed on to their children. As expertise evolves into ever more specialized nested groups of experts, each new generation must also track the constantly evolving landscape of knowledge and skill clusters. A closer look at recent research on how children come to learn from others will provide a framework for considering what might be changing with zoomers and their successors.

In addition to considering different points of view, we often consider different realities. We imagine other possible worlds to better understand our own. While young children show some signs of thinking about such alternatives, impressive growth appears in the preschool and early school years. Children become skilled at pretending that the present differs from what it seems.[2] They learn to think about different possible futures and different possible pasts. They improve their abilities to reason counterfactually and can ask increasingly subtle what-if questions and experience more powerful feelings of both regret and relief. They sharpen their abilities to distinguish the implausible from the impossible. All these skills are brought to bear on their abilities to solve problems and create novel things. After examining how children come to think about other points of view in the present, we'll turn to developmental changes in how children come to think about other possible worlds. In both cases, our goal is to better understand the implications for growing up in a world of rapidly changing technologies.

Explanations and Others

When children want to expand their explanatory understanding, they learn early on that other people can enhance what they learn on their own. But various skills must be developed to fully benefit from the wealth of knowledge around them. Simply being aware that others have knowledge different from one's own is inadequate.

Communicating with Others

How adults talk about the world with children influences how much those children learn and what they choose to learn about. One large-scale study of causal learning about mechanical components examined parent-child interactions in museums across the United States.[3] The study focused on interactions while constructing a device with several gears. Children connected gears together and then set the constructed device spinning. The more parents engaged in causal talk about the exhibit, the more the children used causal language themselves and the more time they spent engaged in exploration of the exhibit. But this relation only held when the parents used causal language during the gear construction phase, not when the child was spinning the gears or after. The second major finding concerned the children's own talk about the exhibit. The more children talked about their potential actions in the exhibit early on, the more they later persisted in playing with the exhibit until they got interesting results. Later talk about the device was not related to persistence.

The language around a child matters in ways that are linked closely to timing. Adults' language anticipating and framing an activity is related to more causal language and exploration by the child. We don't always know when the effects in observational studies of this sort directly arise from hearing causal language or when they arise from a third variable, such as prior experience, but in conjunction with experimental studies, researchers agree that the right kind of language at the right time matters. Specific styles of communication between adults and children are especially effective for learning. A long tradition of studies shows that adults asking open-ended questions prompt more exploration and learning, especially if those questions focus on underlying causal relations.[4] Moreover, very young children learn that some people are better sources of information than others not just because of prior records of trustworthiness and accuracy but also because they give hints about their areas of expertise.[5]

Children learn that different adults have different areas of expertise. They grasp the idea of divisions of cognitive labor and why one might favor advice from one person about how to heal sick animals and from another on how to fix broken machines.[6] They don't always sort experts into the most useful groups. Younger children may focus more on goal-centered categories than on the intrinsic causal structures that give rise to things; thus in searching for someone to heal sick animals, they may initially think of expert healers and only later think of expert biologists. These developmental progressions are hardly surprising given that adults also show shifts in thinking about how clusters of knowledge are distributed into realms of expertise.[7] Adults get more things right than children, but they also can miss the scope of expertise and the many levels involved. Adults also construct areas of illegitimate expertise, such as astrology and psychic abilities.[8]

Information Chains

If a friend makes a surprising claim and refers to three different sources who have confirmed that claim, the claim is more convincing than if the friend refers to just one source. Surely, the larger the number of people who support a claim, the more we should believe it. However, that simple rule can mislead. If the supporting people are not independent but are tightly affiliated, we should be less convinced and perhaps even have greater doubt. If three people who have never met all recommend a particular restaurant, that recommendation is likely to be more reliable than if it came from three roommates. This is only one of many possible examples of the challenges of vetting sources. Most adults realize that a coalition is usually less convincing than three unrelated individuals, but they can overlook other reasons for doubting three separate sources. Even when the three sources don't know each other and come from different backgrounds, they can be unreliable. Sami Yousif and Rosie Aboody in our lab led a series of studies showing how adults can be misled and how children are at even greater risk.

In the first set of studies, adults failed to look further back in the source chain and consider how three sources obtained their information. As a result, they failed to discount claims for which three sources all obtained their information from a prior single source.[9] This is not merely an arcane academic finding. It is ever more common in real life as information spreads rapidly through the internet. As newspapers, broadcast news, and online news operations scale back on investigative reporting, they often simply parrot information retrieved from other sources. For example, "Pizzagate," the malicious false story about US politicians running a pedophile ring out of a pizza shop, was picked up so quickly and transmitted by so many different sources that it was easy to think these were independent news-gathering efforts. Yet they were all depending on a common source:

a person or group who created the fiction for political reasons.[10] A casual glance at a story reported by several news services can easily miss how all their information ultimately originated from one source.

In subsequent developmental studies in our lab, young children had some sense of the greater value of multiple sources, but not until about age six did they start to see the special importance of primary sources. In the same situations, adults easily appreciated the need to determine how the information originated.[11] Younger children's failures have legacy effects in adults. We may know in principle that a provocative news story could seem well grounded when it appears in a wide range of outlets with different reporters but be on thinner ice if all those reporters' sources can be traced back to a single source with vested interests. Yet, in our real lives, we rarely have the time or motivation to trace backward from seemingly diverse sources to find the origins of a story.[12] In those circumstances we can fall prey to the same illusions of consensus as children do.

The quality of information transmission chains is central to the growth of explanatory understanding and insight. Learning explanations from others optimally requires tracking how knowledge is distributed in other minds, how those distributions lead to distinct areas of expertise, and how well grounded they are by chains back to original sources. Simple versions of those tracking skills appear early and can flourish throughout our lives, but their growth depends on expanding causal models of the world. Those models suggest relevant domains of expertise for navigating social networks and for unpacking knowledge transmission chains. Severely impoverished models of how the world works may undermine our ability to track experts and to know where their apparent expertise originated.

Metacognition in Social Contexts

Traditionally, researchers have seen metacognition—the ability to think about thoughts—as slow to develop in young children. In many tasks, thinking about one's own cognitive capacities seems inextricably linked to advanced executive processing skills. The ability to manage cognitive activities, such as inhibiting "gut" responses long enough to reflect on alternatives, does develop during childhood and adolescence and is related to maturation in multiple areas of the frontal lobe. But these growth trends are not the full story about metacognitive development. In social settings, metacognitive skills are evident at much younger ages. This finding fits with a more recent tradition emphasizing the social nature of much higher-level cognition. For example, reasoning skills may be most evident when people disagree and engage in arguments.[13]

More specifically, metacognition may be enhanced by the need to coordinate cognitive resources in social groups.[14] When a group is trying to understand a problem, greater progress is made when each member indicates the level of confidence behind their proposed explanation. Young children may benefit not only when they can use such confidence information to evaluate explanations offered by others but also when they can communicate their level of confidence about their own knowledge and understanding. A child who signals little confidence in their understanding should elicit different responses from adults than when they indicate high confidence. If a child tentatively explains how rainbows form in terms of "something to do with how sunlight gets changed by raindrops," an adult might notice the tentative nature of the remark and immediately act to reinforce the explanation with further details and encouragement. In contrast, if a child explains how a disease is spread in completely incorrect ways but does so with great confidence, an adult is likely to respond differently. All of this requires substantial metacognition.

By embedding a task in a social setting, a recent series of studies found evidence for explicit metacognition in preschoolers.[15] Three- and five-year-olds were asked by others to help them figure out what was in a container. In one condition, the children had complete knowledge of what was in the container. In the other, they only knew about some of the contents. In trying to help the other person, even the three-year-olds spontaneously declared their uncertainty in the partial knowledge condition but not in the full knowledge condition. Such displays of explicit assessments of their own knowledge states have not been routinely found until age seven or later in tasks where children are assessed as individuals removed from a social communicative setting.

The differences in subjects' ages across these tasks suggests that the growth of explanatory understanding may be accelerated when children are embedded in social situations where there are shared goals of helping to solve a problem or achieve consensus through discussion. Social facilitation may be yet another context where mechanism knowledge has broader cognitive consequences for children. Thus, less mechanism knowledge in children may lead to less-developed metacognitive senses of when they have strong and weak understandings. As a result, such children benefit less from more knowledgeable others, a deficit that may snowball over time. As the deficit builds, they might well decide to devote their efforts to learning other sorts of information.

Social embedding can also incur costs related to group affiliation effects. My colleague Yarrow Dunham has shown how surprisingly arbitrary clues to group identity, such as the color of one's shirt, can create in-group versus out-group effects. Information about properties of ingroup members is regarded more positively and leads to more favorable inferences than when the same information is conveyed about an outgroup.[16] The presence of such effects early in the preschool years leads to more impoverished uses of valuable information.

Views of Virtuous Ignorance

Metacognition is related to a more general ability to consider knowledge directly and sense what levels of understanding are feasible for individuals to have in various settings. One aspect of this ability is knowing when declared ignorance indicates a cognitive virtue instead of a cognitive limitation. Jonathan Kominsky led a series of studies in our lab exploring how the notion of virtuous ignorance might develop.[17] The studies examined cases where a person who says they do not know an answer to a question is more knowledgeable about the topic than a person who provides a confident and precise answer. Suppose someone asks exactly how many leaves were on all the trees in Central Park on September 23, 2025, at 9:00 am. A person who confidently declares that there were 10,365,972,128 leaves on trees in Central Park at that time should be seen as less knowledgeable than another person who says they do not know the answer to that question because it is not possible to know that kind of information with any degree of accuracy.

When such scenarios were posed to school children, nine-year-olds judged the person claiming ignorance as more knowledgeable, while five- to eight-year-olds thought the confident person was more knowledgeable. However, the five- to eight-year-olds were able to judge the precise number of leaves as unknowable when it was presented in isolation and no confidence rating was involved. The task became more difficult when younger children had to integrate those two pieces of information. It is not yet known whether the younger children would have performed better if the task had been embedded in another task involving cooperative communication.

The virtuous ignorance studies show how assessments of knowledge can vary in difficulty when combined with other factors such as feasibility. In addition, those studies focused on facts, not explanations. With explanations, feasibility assessments are likely to be more insightful, but also more likely to be undermined by weak mechanistic knowledge.

Another aspect of metacognition concerns intellectual humility, namely accurately appreciating the limits of one's knowledge. Most people tend to overestimate their knowledge. But their estimations also differ as a function of how much they know. The less knowledge people have, the more they tend to overestimate how much they know.[18] Increasing knowledge and understanding reveals all that remains to be known, a phenomenon frequently noted by those with spectacular levels of knowledge. For example, Darwin noted that "ignorance more frequently begets confidence than does knowledge."[19] Maxwell, similarly, said, "Thoroughly conscious ignorance is the prelude to every real advance in science," and Russell wrote that "in the modern world the stupid are cocksure while the intelligent are full of doubt."[20] As knowledge expands, especially explanatory knowledge, the enormous terrain of what is still unknown comes into much sharper relief. If people have a natural talent for discerning causal mechanistic patterns, thwarting the ability to build up that knowledge impairs the ability to be appropriately humble.

Parent-child interactions can influence children's intellectual humility. In one study led by the psychologist Candice Mills, a subset of parents provided incorrect answers to test items without indicating knowledge gaps. The children of those parents rated their own knowledge especially high in comparison to other children. In that sense, parents showing less humility in their explanations had children who were also less humble in assessing their knowledge.[21] This embedding of humility indicators in natural conversations suggested an influence, while more direct self-assigned ratings of knowledge were less revealing. For both adults and children, assessments closer to real-life social situations suggested how parents with greater knowledge might instill intellectual humility in their children.

Young experts with corresponding knowledge about their knowledge, or metaknowledge, can be found in many naturalistic settings. For example, I recently met two young experts on freshwater ponds. One day in early June, two boys (ages six and eight) from our neighborhood showed up at our door asking if they could fish in our half-acre pond. I said that no one had ever caught any fish in our pond and I wasn't sure there were any fish. They looked at each other and grinned and asked if they might at least try. Within ten minutes they hauled out a five-pound largemouth bass. As we watched them catch and release new fish every few minutes, their mastery of pond fishing became obvious. When I asked how they caught the fish, I was flooded with sophisticated knowledge about pond ecologies and fishing, including elaborate metacognitive observations by both children. For example, they described at length situations where they still had gaps in understanding and how they were trying to remedy those gaps.

Future studies will provide guidance on how metacognition can be enhanced in younger children by embedding such judgments in socially compelling tasks, such as when a common goal is shared by a group and requires integration of diverse skills. Those enhanced metacognitive skills may be crucial in a world of ever more unexplainable things. More refined senses of areas where we have little or no understanding may not give direct insights, but those senses can support our abilities to develop strategies for coping with what remains unknown.

Other Worlds: Counterfactuals

We have seen how counterfactual thought in adults is intimately involved in constructing and understanding causal explanations. However, choosing the best counterfactual for the topic at hand is not always straightforward. Recent research on the development of counterfactual thought provides new ideas on how people of all ages might use counterfactuals most effectively.

Assessing counterfactual thought in young children and infants imposes special challenges arising from their more limited language skills. Because many ways of talking about counterfactuals can baffle schoolchildren, classic developmental theories considered true counterfactual thought to appear only when children enter early adolescence.[22] However, as researchers have more closely considered how counterfactual thought might manifest in the mind, they have found it at much earlier ages.

A collection of researchers from our group and five other labs addressed the difficult issue of whether young children are qualitatively different from older children in their attempts to engage in counterfactual thought.[23] We found that, at least by age four, when a simple set of mechanical collisions were presented to them, children were able to engage in true counterfactual simulations in which they systematically thought of alternatives in order to make predictions. They often simulated different alternatives than most adults would, but those differences are more suggestive of a gradual pattern of developmental change than a radical shift in representational competence. Young children seem to be capable of counterfactual thought and appreciate its value. But because of their weaker background knowledge, greater cognitive limitations in representing and talking about multiple alternatives at the same time, and tendency to put more weight on certain salient alternative simulations, preschoolers make different judgments.

These results have broader implications for how adults engage in counterfactual thought, how novices might differ from experts in what counterfactuals they pose, and how we might most effectively use counterfactuals to make sense of predictions made by AI systems. With adults and novices, classic studies on expertise show that experts in physics sort physics problems into different categories than novices.[24] For example, novices tend to sort mechanical situations into categories such as ramp problems, pulley problems, and collision problems, while experts might sort them into categories corresponding to Newton's laws of motion. It seems likely that novices would also pose different counterfactuals in thinking about such problems because their underlying similarity spaces are so different.

Consider the finding that five-year-old chess whizzes remember midgame configurations of chess pieces far better than adult novices.[25] Their greater ability to store and recall such large patterns implies that they would be able to pose more and better counterfactuals when thinking about chess problems than same-age peers who were only beginners. This difference seems almost inevitable, as it is difficult to imagine playing chess without constantly entertaining what-if scenarios. It would follow that children with extensive exposure to working mechanisms will more easily generate relevant counterfactuals for mechanical devices in comparison to those with impoverished exposures. Just as with metacognition, extensive domain knowledge can support the ability to think about other possible situations in that domain. In chapter 9 we will see how this may have implications for using counterfactuals to better understand AI recommendations.

Expertise in mechanical systems is not the only way to facilitate counterfactual thought in young children. Various studies have argued for early counterfactual thought in the realm of reasoning about behavior and other people's minds.[26] Given arguments about psychology and physics being especially robust early domains of thought, rich experiences in both these areas may be especially central to promoting both metacognition and counterfactual thought.

Creating New Things

Children go beyond their normal daily routines by pursuing novel experiences. Toddlers become adventurous explorers as soon as they can walk with some proficiency. Walking may be more cognitively expanding than we normally appreciate. It expands visual experience by enabling the walker to more easily see off to the horizon in all directions, unlike the default downward-looking behavior of crawling.[27] Walking also enables much longer explorations over more varied terrain as children can be drawn more naturally to far-off things.

In terms of the objects around them, young children fiddle with things on their own before they can speak. Classical studies by Piaget and others describe infants trying out different actions on objects and sometimes delighting in the unexpected consequences of those actions. A child might be playing with a toy in a traditional manner but then discover new effects that happen when that toy is used in an atypical way. For example, a two-year-old may discover that letting a toy truck go backward down a ramp makes it swerve and crash in exciting new ways. Younger children may stumble on these new actions by accident and then seek to reproduce them rather than planning them in advance after reflection. Other times, after stumbling on a new effect that older children find interesting, younger children ignore it and go back to the standard use. In such cases they resemble hidebound traditionalists.[28]

Younger children almost never figure out all the possible novel actions to take on a toy and then explore them all systematically; but few adults do this either. Indeed, many product liability cases pivot over whether a manufacturer should have reasonably foreseen an atypical use of a product that caused a serious injury.[29] Children are especially limited in this way, as many parents acknowledge when asking, in exasperation, something like "How could you not see that tying our dog to your bike while riding it was just asking for trouble?" But again, children aren't completely unable to think about alternative uses; they are just more limited.

In this context, the "alternative uses" test is notable as one of the classic tests of creativity. Even if we acknowledge controversies over the predictive value of most creativity tests, children's more limited abilities to generate alternative uses do fit with recent claims that they have less well-developed skills in innovating and imagining. Despite having precocious abilities to acquire and refine causal explanations, young children can show dramatic deficits in their ability to come up with new solutions to problems. The magnitude of these failures is a relatively recent discovery.[30] We've seen that children often uncover and detect information about mechanisms, including through indirect means such as observing the diversity of actions a device exhibits. But these eager and adept learners often fail to create simple tools to solve new problems.

Across several tasks, younger school-age children show fewer innovations than older children and adults.[31] One of the most common innovation tasks invites children to retrieve a small bucket with an upright handle from a deep tube. The solution involves bending a strip of metal (or a pipe cleaner) into a hook and using the hook to retrieve the bucket by its handle (see figure 6.2). While even some species of crows produce such tools and successfully retrieve the container, young children usually fail. The psychologist Andrew Shtulman has described this developmental challenge as the problem of "learning to imagine."[32] Because these failures have been found in a wide range of cultures, they are not seen as a product of specific cultural biases toward thinking creatively.

To understand what is happening in such failures, we need to take a closer look at the typical methods used to assess innovation. The tasks often resemble the classic Kohler insight problems with apes. One example is the two stick problem. A chimpanzee was given two bamboo sticks of different diameters in a setting where a desirable food was positioned out of reach of either stick. After several failed attempts, the chimpanzee realized that the problem could be solved by pushing the thinner stick into the end of the thicker one, producing a new sufficiently long tool. Given the setting and the objects available, it was the only solution to the problem. Yet real-world innovation challenges are rarely problems of finding the one right answer, and it is unclear how one skill maps onto the other.

Many explanations have been proposed for the innovation failures observed in younger children. Children might be hobbled by inadequate executive processing skills, greater difficulties with having foresight and

Figure 6.2
Failing to innovate as well as a crow. Researchers were initially stunned to find crows devising new bent-wire tools to solve a problem that most young humans failed to solve. *Source*: author.

generating counterfactuals, weaker analogical reasoning skills, or constraints posed by tasks where there is only one solution. All of these plausibly contribute to poor performance; but all might be mitigated in tasks where children could bring to bear more elaborated background knowledge.

From a different perspective, innovation failures might also be a side effect of copying how others manipulate objects to attain goals. Derek Lyons and I used the term *overimitation* to describe humans' tendency to imitate causally implausible components of an observed action.[33] For example, an actor might remove a toy from a puzzle box by performing some actions that are clearly related to getting inside the box and also performing other actions that seem unnecessary. The phenomenon was initially discovered in studies comparing apes with children where only the children overimitated (i.e., mimicked the unnecessary as well as the required actions).[34] Despite its apparent irrationality, engaging in this kind of behavior can be a useful strategy for learning how to operate devices before one fully understands how they work. Overimitation depends on the assumption that people do not perform unnecessary actions to operate a device.[35] Thus, even if a step appears implausible, one should copy it unless it appears to be blatantly impossible. Overimitation may be a uniquely human behavior. My colleague Laurie Santos has argued that dogs, while adapted to be highly sensitive observers of human behaviors, do not engage in overimitation, but largely ignore seemingly irrelevant causal actions.[36]

These findings have been interpreted by some as showing that people don't need to understand how a thing works for its use to be transmitted to later generations. They just need to copy what the culture does at large.[37] In many cases, however, overimitation supports causal reasoning and innovation. The tendency to copy and remember seemingly useless manipulations of objects' parts may yield downstream benefits when a more subtle utility of a previously mysterious action is discovered or when some of those seemingly silly manipulations suggest a novel solution to a problem. For example, you might observe drivers in some European countries performing the odd and seemingly senseless action of opening their door with their right hands only (for cars with the driver's seat on the left). Only later might you realize that this action was introduced to ensure that drivers noticed bicyclists who might be passing their parked car in an adjacent bike lane.

All these factors may well be part of a full explanation of early failures to come up with novel solutions to problems; but they do not mean that young children are unable to innovate. Young children are surely more limited in this ability, but given an appropriate context, task, and background experience, even preschoolers may create novel tools and other solutions to problems. If asked to make shelters for toy animals in a setting full of many objects and materials, young preschoolers will easily come up with novel constructions. Prior studies may have underestimated the importance of experience by using tasks where a brief training episode in a novel setting is meant to control for background knowledge. In contrast, children may need to gradually build up relevant knowledge over months and years. This typically occurs through immersion in a culture with the appropriate technology and the right kinds of interactions with that technology. Those long-term experiences may enable children to build up conceptual tool kits of easily deployable units that they can bring to bear on innovation tasks.

Years of watching how others fix and modify things help children assemble a reliable set of innovation building blocks. Certain general classes of modifications may be abstracted away from extensive experience: shape changes that make stronger pullers and connectors, ways of positioning elongated objects to increase leverage, and ways of bracing to enhance rigidity, among many others. Well-developed and practiced knowledge can catapult children ahead on tasks that would normally seem too hard for them. As noted earlier, five-year-old chess wizards show remarkable memories for locations of chess pieces while performing just as badly as other five-year-olds on memories of displays of other objects. But when the practiced element is a more fundamental mechanical understanding that applies to a wide range of artifacts, more extensive exposure to just a few of those devices may have especially broad positive consequences for innovation.

Cognitive tool kits that build on mechanistic patterns are particularly relevant to the changing world of technology. They also fit well with young children's inclinations to think about goal-driven results, as seen in a set of studies led by Judith Danovitch in our research group.[38] While children as young as five were sensitive to how knowledge could be clustered roughly according to traditional science disciplines, five-year-olds were more strongly inclined to structure knowledge according to goals. Thus they might find it quite easy to infer that a person who knew all about how to catch fish would have other pieces of knowledge that would support that goal (e.g., knowing what fish like to eat and what time of day they are the hungriest). For discipline-driven clusters, they might also infer that a person who knew all about how fish have babies would know more about how fish pump blood through their bodies or how fish digest food. But such discipline-based cluster inferences were less obvious to younger children than goal-based ones.

Researchers often compare children to little scientists; this comparison can be both useful and misleading. In some cases, viewing children as little engineers may more accurately characterize what they are doing: finding goal-driven solutions to practical problems and organizing expertise areas around goals.[39]

The developmental pattern here roughly reenacts the history of human areas of expertise. In most areas, intuitions about goal-related expertise preceded those about discipline-centered expertise. The first areas of expertise with respect to biology were likely centered around knowledge clusters concerned with hunting animals and healing sick humans and animals. As experts in these goal-driven areas became more sophisticated, and as they looked for explanations of underlying causes, they started to converge on deeper principles concerning such topics as physiology and metabolism. The same story recurs for many other analogous emergences of sciences out of goal-driven expertise, such as cooking begetting chemistry and shelter building spawning physics.

Tinkering

As part of inventing new things, tinkering can be helpful. Here we consider tinkering as a hands-on activity in which children explore and manipulate properties of an artifact with the goal of changing it or getting it to do new things. This typically occurs through iterations of trial and error. The idea of tinkering can be extended to melodies, poems, and other nonphysical creations, but our focus is on physical artifacts. Finally, tinkering is thought of as a playful, intrinsically enjoyable activity.

The earliest origins of tinkering depend on how loosely one interprets an infant's goals and how sustained the engagement must be to count as a tinkering event. Tinkering plausibly occurs by twelve months of age, when infants explore the consequences of banging on an object in different ways or experiment with how one object stacks on another. Tinkering then increases among toddlers and preschoolers as they learn more about the properties of objects. A toddler might add objects to a floating cup in a bathtub to see what makes the cup sink, or they might see what causes a tower of objects to be unstable.

While tinkering has been long regarded as a core activity by advocates of maker movements, only recently has it been examined more carefully in terms of how it might support innovation.[40] Interest has soared in response to multiple reports of innovation failures in younger children. Most adults, including parents and schoolteachers, believe children are creative and imaginative; indeed, they will often describe children as more imaginative and creative than adults. Evidence supporting such views includes cases where children coin new words and metaphors (e.g., calling a car's tires its shoes). In other instances, they engage in play scenarios that aren't bound to adult conventions. When our children were young, we sometimes found them flying rocket ships with caveman pilots or building a small fort in our yard from recycling materials in a nearby bin. But critics suggest that such examples are cases where lack of knowledge is masquerading as creativity. The children aren't saying "wouldn't it be clever and different to have a caveman pilot a rocket ship." They simply don't know enough about the time frames for cavemen and space travel technology.

Tinkering is an especially good way to promote innovation. In a recent study, the more children of any age tinkered, the more they could innovate in tasks requiring novel solutions.[41] Children between four and twelve years were presented with a wide variety of possible materials they could use to make tools to retrieve prizes from a box. As in prior studies, younger children showed less innovation. However, when age was controlled for, investigators found that the more children "tinkered," the more they innovated. Tinkering was defined as keeping a greater proportion of materials from prior failed attempts to use again in later attempts and adding more new materials to a tool after a failed attempt.

In the tinkering study, the innovation problems were different and arguably more naturalistic than those in prior studies. Many solutions were possible; there was not just one right solution. The tasks were more open-ended and there was no time pressure. Children could also engage in repeated attempts. These differences support greater tinkering as well as more innovation arising from knowledge gained through tinkering.

How helpful could just the right prior experience for a particular innovation task be? Younger children's markedly lower innovation rates across many different cultures would seem to argue against effects of different experiences; a closer look at such studies, however, suggests an alternative conclusion. The largest cross-cultural study interviewed children from four non-Western cultures and one Westernized culture (people in Brisbane, Australia).[42] In three tasks, children from all five cultures innovated more with increasing age (three to twelve years). But there was also a large group difference, with more innovation in Westernized children of all ages than in those from all other cultures, who all performed similarly. The authors were unsure why the Westernized children showed so much more innovation. They speculated that Westernized children have more exposure to a wider variety of tools and tool-like toys that invite construction and modification, and they cited several other studies showing such analogous differences.[43] It seems likely that the three tasks used nicely aligned with the artifacts that Westernized children encountered as tools and toys. Presumably, if children were assessed with tasks more familiar to non-Westernized cultures, the innovation success patterns would have been reversed. If that conjecture proves correct, it suggests that innovation is strongly influenced by prior experiences and the amount of relevant background knowledge brought to a task. This increases the plausibility of generational differences as well.

Tinkering may provide more effective support for innovation when it is done by a group. Each member of the group can observe many more possible manipulations and modifications than they would generate on their own and thereby be able to come up with more effective solutions. This is often a central argument for having maker movement communities rather than individuals working in isolation.[44]

Tinkering is also a form of risk-taking. We can break or make a tool worse, sometimes in ways that cannot be reversed, by tinkering with it. Many of us can remember trying to fix a device as children only to break it irreparably. Suppose recent alarms about over-parenting and excessive worries about children's safety are warranted. If so, children of such parents may be tinkering far less than they used to, not only because today's things have fewer parts to tinker with, but also because parents may be keeping children away from older, more tinkerable, real-world devices out of safety concerns.[45] Prowling around junkyards and boatyards was a favorite pastime for many children in my generation. Such excursions are less common today.

* * *

Explanatory understanding soars early on in social settings. Children learn as members of knowledge communities. Their views of explanation quality and believability are shaped by how they think about the minds of others and how they integrate that information with the situation at hand as well as with relevant histories. They do all this intuitively, long before they start formal schooling. They also show impressive abilities to adapt to changes in knowledge communities. In some cultures, these early skills may decay when they are not valued. As the explainability of the world around us changes, supporting and nurturing children's explanatory understanding becomes ever more important.

Children are rarely lone agents building understandings on their own in an unchanging world. They rely on and evaluate others, think about alternative realities and futures, and actively change and manipulate things. They can create new things as well, but their ways of innovating are more constrained. Most adults also grossly misconceive the pace of innovation and how much it depends on communities; this holds for innovation in light bulbs, radios, FM radio, TV, transistors, solid state circuits, fiber optics, and even the lowly sewing needle. When adults misunderstand the dynamics of innovation in five- to eight-year-olds so badly, they are ill-equipped to foster it in their children.

In the past two chapters we have seen an early emerging drive to go beyond the obvious—to dig deeper inside things while at the same time learning how to rely more and more on other minds and how to ask what-if questions. While children tend to explicitly underestimate their reliance on other minds, they have considerable skills in outsourcing. To what extent are both processes dependent on the nature of the physical world, especially artifacts? The case for the child's drive to discover working innards is apparent; it may apply to outward-looking connections as well. We may all heavily rely on artifacts around us to evaluate knowledge in other minds. We can observe how others use, modify, fix, and even create artifacts; and we can ask how they use that information to learn how they think and what they know. We can also do much of this by simply examining an artifact in isolation. But, as we lose relevant knowledge about how things work, we will also lose some of our ability to map the terrain of expertise around us.

A child's experiences with devices today seem to be different from that of a few decades ago. Given that young children have a growing drive to explore and discover hidden causes and relations, and given that they can detect such information at highly abstract levels and retain it over time, does it matter that today's children no longer have as much access to this information? The innovation differences across cultures suggest that experience does matter.

If today's children received impoverished information about inside parts and relations, how consequential is that ignorance in comparison to other benefits? Is it compensated by new streams of insight not available to earlier generations? Is the ease of accessing information on the internet making the deficit moot? Because most of the demonstrations of children failing to innovate have been done over the last two decades, it leaves open the possibility that children fifty years ago would have innovated more in the same tasks.

We've seen how both as individuals acting in the here and now (chapter 5) and in interactions with other people (this chapter), children's ways of thinking may be influenced by changes in exposure to classical mechanisms. Some possible consequences may accelerate in adolescence and young adulthood and be most evident in this age group. In chapters 7 and 8, we turn to consider generational effects and how they might continue beyond childhood.

III Downstream

7 Construals in Cohorts

In October 2023 the columnist Pamela Paul wrote a piece for the *New York Times* deploring the decline of home economics and shop classes throughout the United States over the previous decades. For Paul, something important was lost when hands-on experiences of real-life relevance were dropped from middle school and high school curricula.[1] She described the intrinsic value that comes with making physical products and how those course experiences also enabled students to fix and repair things, thereby giving them more power over their lives and less passivity. Her essay elicited a flood of responses in the comments section (over twelve hundred in less than twenty-four hours). The responses could be clustered into several categories. Most of the responses also saw the decline in shop classes and home economics as regrettable. Several readers in that group described cases where zoomers were surprisingly helpless in changing a car tire, putting a bike chain back on the sprockets, fixing a simple lamp, or even doing their own laundry.[2] People in this group also described how shop classes prepared them for careers as scientists and engineers. The positive responses also extolled the cost savings arising from being able to do their own work and repairs.

The remaining relatively small set of critical comments covered diverse reactions. Some mentioned experiences with terrible teachers or problems relating to gender biases. Others felt such courses should focus on other topics such as modern financial practices. A surprising number expressed concerns that the tools and appliances used in those classes would open schools to lawsuits. (It was not clear why lawsuits were not a problem a few decades earlier). Finally, a small group forcibly argued that dropping such classes was exactly what was needed to free up the minds of youths for mastering the avalanche of new digital technologies. Arguably, the small minority of critical responses simply reflect the aging demographics of readers of the *New York Times*, especially those who read the columns. But if negative views of future generations' cognitive repertoires are simply alarmist speculations by boomers, those with positive views need to make their case. They should provide detailed examples of how new cognitive capacities are emerging; and they need to explain how all of us might support those new learning trajectories as part of our contribution to cumulative cultural evolution. They might explain, for example, how whatever was lost by closing shop classrooms was more than compensated for by replacing them with computer classrooms (figure 7.1).

How feasible is the proposal that declining exposures to mechanical systems could have lifelong influences on how we think, feel, and behave? To get a better sense whether major consequences are feasible, it is useful to briefly consider broader claims about how newer "digital media" have changed us. Can significant shifts happen in just a decade or two; and, if shifts do occur, are they constrained to specific spheres of mental activity such as emotions?

The Calculator Crisis That Never Was

Over a five-year period starting in the late 1970s, people throughout the world shifted in how they did math. With the introduction of inexpensive

Figure 7.1
Shop classroom versus computer classroom. These two school settings would seem to provide different experiences in terms of how children interact with the class materials and each other. *Source*: author.

portable calculators, they off-loaded more and more calculations to those devices. Many people in my generation were stunned to see teenage store clerks turning to calculators to do simple math calculations that their predecessors easily did in their heads. It was not uncommon to see the same store clerks panic when their calculator batteries died. Teachers and parents worried that calculators would destroy math calculation abilities and quantitative intuitions. Yet no large-scale negative consequences were found.[3] In many ways, the absence of significant effects is surprising. If calculator-reliant children and teens were doing far less math in their heads, shouldn't they lose some intuitive sense of how magnitudes can change when multiplied by a certain amount? Perhaps any losses of mental storage of addition and multiplication facts were compensated by a greater sense of the approximate values of a much larger range of computations. The introduction of a new technology clearly changed the degree to which people did math calculations in their heads, but no differences in math abilities or intuitions were apparent. Perhaps the right studies weren't conducted—but the absence of reliable effects suggests we should use caution when predicting how technological changes might influence how we think.

Digital Downers and New Media Boosters

Observers of the changes in media over the past few decades are polarized about whether those changes have positive or negative effects. When effects are discussed, they are most often described in tweens, teens, and young adults. The effects may start earlier in childhood, but the discussions center on older youth. This emphasis on adolescence and young adulthood reflects a view that generational differences in that age group are likely to be more stable and long-lasting in the decades that follow.

The Case Against New Digital Media

Far fewer people read books now than did a few decades ago. Declines in book reading have occurred in all age groups, across genders, and in all levels of education. Reading extended text may be displaced by time spent browsing social media, short-form videos, and brief paragraphs or phrases. For children between nine and seventeen years of age, reading levels are the lowest in over forty years. More than half of all eighth graders use social media more than three hours a day, and the mean time spent on screens of all sorts is eight and a half hours per day.[4] A 2024 essay in the *New York Times* argued for paying a tween one hundred dollars to read a book. The author of the essay also noted that when she paid her child in this way, that tween requested, and read, the sequel for no fee. The several hundred comments were sharply divided about whether paying your child to read was a good idea.[5]

The daily information diets of older children and adolescents are different from just a few decades earlier. Almost half of all ten-year-olds in the United States owned their own cell phone in 2022; a strong majority owned phones only a year or two later. By most estimates, a large majority of the world's population either has a cell phone or easy access to one. Cell phones, especially continuously evolving smartphone versions, enable constant access to social media and video games. Gaming and media interactions are continuously becoming more elaborate and immersive on tablets, game consoles, and computers.

This new communication landscape is of concern to a group we might call the *digital downers*. The English scholar Mark Bauerlein, author of two books, both with titles starting with the phrase "the dumbest generation," has repeatedly described the rise of digital media as having massive negative effects on general cognitive skills.[6] Similarly, the writer Nicholas Carr worried that new digital media are pushing today's youth toward an intellectual shallows:

> Dozens of studies by psychologists, neurobiologists, educators, and Web designers point to the same conclusion: when we go online, we enter an environment that promotes cursory reading, hurried and distracted thinking, and superficial learning. . . . As supple as it is subtle, it's been the imaginative mind of the Renaissance, the rational mind of the Enlightenment, the inventive mind of the Industrial Revolution, even the subversive mind of Modernism. It may soon be yesterday's mind.[7]

From a different perspective, focusing on historical changes in knowledge, the writer Simon Winchester worried that we may be headed toward an "intellectual apocalypse" in which just those cognitive tasks that we evolved to handle will be taken over by machines, leaving us with superficial processing and little more.[8]

These examples are part of a large pessimistic chorus concerned about declines in general cognitive skills relating to attention, memory, and various versions of deep reflective thought.[9] Those views tend to emphasize general skills and not problems with specific kinds of content, such as mechanistic models. In one sense, it is trivially true that the disappearance of some outdated devices will cause local knowledge loss; however, that hardly seems to be a source of concern. Why should we want to remember how to use punch cards to input programming instructions to computers? It would seem maladaptive to clutter our minds with such details. But if a disappearing aspect of technology gave us insights into more general principles and models, the loss might matter.

The most dramatic and arguably best documented claims of negative effects of new technologies are found in social and emotional development and in the mental health of teens and young adults. The psychologist Jean Twenge reports strong correlations between negative emotions and time spent immersed in social media websites and similar platforms.[10] More specifically, Twenge describes how major rises in anxiety, depression, and loneliness have been found in several large samples of zoomers, especially when they engaged in heavy social media use between the ages of thirteen and fifteen.[11] Most of the evidence is correlational and possibly due to other variables.[12] However, some experimental studies have found that reducing digital content reduces depression and anxiety.[13] The psychologist John Haidt has argued that over two dozen experimental studies now support such effects.[14] If future studies continue to unambiguously show that the negative effects are primarily direct consequences of time spent on social media, those studies would be the strongest demonstrations of new technologies negatively impacting younger generations.

A different set of worries ask whether newer social media and constant texting and messaging induce a kind of learned distractibility. With information coming in on so many channels, often at the same time, and constantly interrupting other cognitive activities, perhaps children are becoming adept at switching between tasks at the expense of sustaining reflective and critical thought. In terms of possible costs, one line of work suggests that multitasking ability does indeed require trade-offs. Extensive multitasking activity can impair memory encoding of items presented while multitasking. Some modest long-term effects on cognitive control and attention have been reported, but none of this appears to have any effect on students' grades. People who are interrupted in a complex task by another task that draws on similar cognitive activities, such as reading, interpreting, and remembering complex screen messages, do worse than those who are not bombarded in this way.[15] The failure to find consequences in classroom performance is surprising, but may reflect what happens in most schools more than any effects of declining abilities related to cognitive control and attention.

The Case for New Digital Media

Citing the many historical examples of cognitive luddites being wrong in predicting the negative effects of everything from writing to television, a group of new media boosters has celebrated the digital future. The science writer Steven Johnson was one of the first to embrace the new media culture in his 2005 book, *Everything Bad Is Good for You: How Today's Popular Culture Is Actually Making Us Smarter.* With discussions of new media formats ranging from the shift to multi-story line television shows to video games, Johnson's book praised the cognitive benefits of contemporary culture. Johnson was not timid in his assertions: "The television shows and video games and movies . . . are more complex and nuanced than the shows and games that preceded them. [This new content] . . . is the single most important new force altering the mental development of young people today, and I believe it is largely a force for good: enhancing our cognitive faculties, not dumbing them down."[16]

In support of his account, Johnson relied heavily on a discovery concerning the history of intelligence test scores—the Flynn effect. Named after the late New Zealand philosopher and social scientist James Flynn, the effect was a continuous rise in IQ test scores from the 1940s to the early 2000s. Flynn discovered that scores over the years were not equivalent, but when adjusted, revealed yearly rises everywhere in the world. By some accounts, members of the "greatest generation," those who fought in World War II, had IQs averaging around 80 in comparison to the average score in 2008 of 100. Flynn later wrote that such a rise in scores was a clear condemnation of IQ tests as measures of real intelligence. No one regarded most young adults in the 1940s as mentally challenged. Others, such as Johnson, speculated that the rise was genuine and should be seen as evidence for a new kind of intelligence tailored for the digital future. Johnson's claim is provocative, but it has become problematic given that IQ scores are no longer rising. An extensive review in 2023 found consistent drops in young adults in the United States in tests from 2006 to 2018.[17] If the new narrative structures of modern media are having positive cognitive influences, the evidence is not found in intelligence test scores.

A wide variety of popular books, mostly not written by researchers, echo Johnson's views. The titles ring of boosterism: *The Game Believes in You: How Digital Play Can Make Our Kids Smarter* and *Reality Is Broken: Why Games Make Us Better and How They Can Change the World.*[18] Ever since the first video games appeared, proponents have celebrated them as powerful new tools for advancing cognitive growth. The arguments can be compelling when buttressed by individual examples of how a child addicted to a game learned enormous amounts about history, biology, or urban planning. Those individual cases surely do exist, but they do not amount to a net positive effect for younger generations. The video game industry's immense profits depend more on cleverly designed reinforcement protocols than on the intellectual growth of gamers.

In short, with respect to cognitive skills pertaining to memory, attention, and executive processing, few consistently strong negative or positive effects of new media have been found. Both sides of the debate can point to correlations between activities associated with new media and with basic cognitive skills; it is more difficult, however, to go beyond those correlations to demonstrate convincing causal effects. With respect to social and emotional factors and mental health, stronger correlations have been observed, and some experimental evidence supports negative causal claims. But, even there, ruling out other potential causes is difficult.

Considering Content

Given the positive correlations between new media use and negative social and emotional states, might any such effects subsequently influence learning of specific kinds of cognitive content? One effect would follow from Jon Haidt's argument that social media exposure increases the incidence of over-parenting.[19] Together, those factors could lead to well-documented declines in free play and risk-taking activities. Those declines could in turn result in diminished personal initiative and greater timidity. Some of these declines have been tied to drops in tinkering activities. Similarly, a rise of tightly scheduled and closely managed out-of-school activities might reduce time spent in unfettered tinkering and exploring. Heavy immersion in the most popular social media sites is likely to also result in less learning of underlying mechanisms.

The generational differences in childhood experiences seem obvious if we consider a large collection of brief articles for children first introduced in *Popular Mechanics* in 1913. These pieces were subsequently published over the next decade in four large volumes titled *The Boy Mechanic* (books 1–4). Setting aside the unfortunate gender bias of the title, it is clear that these volumes were treasured by earlier generations. Children used scrap materials to construct their own versions of devices selected from the thousands of illustrated projects. The illustrations were not step-by-step like IKEA assembly diagrams. They were line drawings with just enough detail that a child could, with practice, figure out the missing steps and parts. The books are still available, and the online reviews are sharply divided. Some praise the lack of step-by-step details and how the illustrations provide just enough information for a child to figure out how to do the rest with the materials they have available and their specific goals. Others see the books as frustrating, even worthless, and describe their children as baffled by the sketchy descriptions. The positive reviews report the books are joyfully embraced by their children, who often are in the seven to nine age range. The negative reviews not only describe the book as informationally inadequate for their pre-tweens and teens; they also describe the activities as no fun for their children and "lame."

Figure 7.2 shows two examples. The self-closing gate has just one moving part and merely depicts a clever idea. Building a solid working gate requires experimenting with different angles of the two support cables (chains) and adjusting clearances and support points on the gate. In doing so, a child could gain a great deal of knowledge about torque, bracing, forces exerted on objects, and the like. The snowball thrower has several moving parts and illustrates how tools are adapted to niches (e.g., snowball size and range). Building a snowball thrower enhances knowledge about leverage, fasteners, cams, springs, and catches. It also reveals how to make functionally similar devices. A nine-year-old leafing through *The Boy Mechanic* might quickly realize how the snowball thrower could be modified to make a baseball pitching machine. After building a dozen or so different devices, a child would have a rudimentary cognitive tool kit containing knowledge of basic mechanical principles and the typical functional components found in mechanical systems.

The books embody an unusual combination of improvisation, frugality, and community involvement. Advice is given on how to secure needed

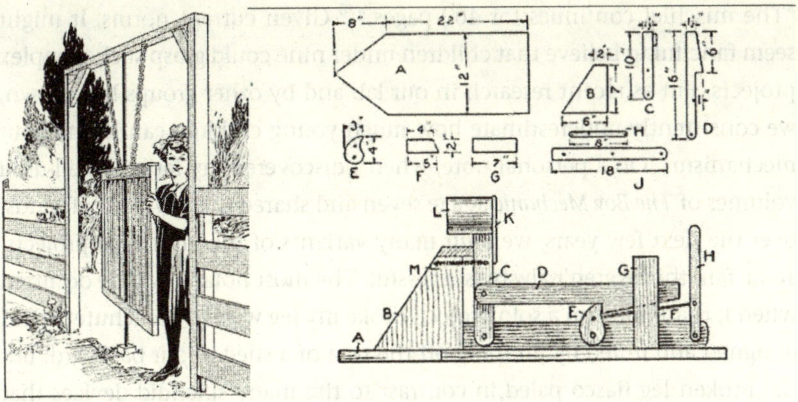

Figure 7.2
A self-closing gate and a snowball thrower. Two of the seven hundred projects depicted in Windsor, H. H., ed. (1913). *The Boy Mechanic* (Vol. 2, pp. 47 & 211). Popular Mechanics.

components from abandoned bicycles, broken barrels, discarded plumbing pipes, and the like. Readers are encouraged to ask for discarded components from their neighborhood merchants, such as broken sewing needles, worn out tires, and defunct agricultural equipment. Many projects contain advice on how children might find it easier to build that project with friends. However, help from adults was almost never mentioned, leaving the strong impression children should try to figure out things on their own, either individually or in groups of peers. Because the directions and illustrations are not exhaustive, they presume a basic mechanical mental tool kit. In contrast, many modern construction toys try to take children through every step in a unique construction sequence with many parts that are also unique. In doing so, they may rob children of creative agency.

Most of the *Boy Mechanic* projects still seem feasible to accomplish today, with some notable cautions. Several projects encourage children to ask their friendly neighborhood blacksmith to fashion a small metal fitting in their spare moments—not a likely resource in the twenty-first century. Other projects, such as working cannons, biplane gliders to carry their builders off steep hills, radium-infused displays, and chemical mixtures containing arsenic, mercury and toxic acids, are unlikely to be approved by even the most permissive parents. In an enthusiastic description of the exciting projects in volume 1, one historian noted with barely disguised glee,

"The mischief continues for 460 pages."[20] Given current norms, it might seem fanciful to believe that children under nine could grasp such complex projects. But as recent research in our lab and by other groups has shown, we consistently underestimate how much young children can learn about mechanisms. On a personal note, when I discovered my father's well-used volumes of *The Boy Mechanic* at age seven and shared them with my friends, over the next few years, we built many variants of the illustrated projects. To be fair, they weren't always successful. The most notable failure occurred when I, thankfully on a solo project, broke my leg with a "parachute" that I designed and tested by jumping off the roof of a shed in our backyard. But the broken leg fiasco paled in contrast to the many splendid devices that populated the houses and yards of my circle of friends.

These youth-centered projects have disappeared in the last few decades. No widely popular counterparts exist for children and young teens today. The recent maker movement might be construed as a revival of this approach, but it differs in critical respects. It has heavy adult involvement, it has often been commercialized with expensive kits sold to schools and other groups, and the projects are much more variable in terms of supporting insight into basic mechanisms. It does emphasize learning through doing and often endorses a grassroots spirit; but it includes everything from weaving to coding to cinema. Perhaps this movement will gradually morph into programs that adopt a more consistently frugal/reuse attitude and help build up concepts of basic mechanisms. But that trend is not obvious today. For example, 3-D printers are often touted as the key tool at Maker Movement centers even as most users have no idea of how the 3D printer works and could not afford one of their own.

Other craft-centered activities have also declined. Throughout elementary school, middle school, and much of high school, students of earlier generations had hands-on experience with concrete models and mechanisms. Outside school, most children today rarely engage in the kinds of activities depicted in *The Boy Mechanic*. Notable exceptions do occur in clubs where older children construct complex devices, but these children are only a small percent of the broader population of their peers. Perhaps tiny clusters of mechanism addicts are all a culture needs—but sequestering expertise and leaving the general population with little or no knowledge of basic causal patterns requires a closer look at potential costs and benefits.

As parts of the young child's world are becoming less explainable, do those differences in early experiences gradually build to have cumulative effects on thought in older children, adolescents, and adults? Is that why adolescents seem so much less interested in underlying mechanisms? Conversations about mechanisms with people in their late teens and early twenties and with people in their late sixties and early seventies seem to have different qualities. Discussions about parts of devices where the mechanisms have remained roughly similar for the past fifty years, such as car suspensions, water pumps, and audio speakers, can expose much larger knowledge gaps in the younger generation.

If such differences are reliably found, perhaps the younger generation has simply adapted to current needs and acquired greater understandings of newer classes of more currently useful causal relations. We have seen from recent and current research that, when mechanisms are easily accessible, young children can show surges of spontaneous discoveries and insights. It follows that when those mechanisms are obscured early on, the effects of reduced exposure to how things work could result in shifting cognitive priorities and interests in later childhood and young adulthood. As a result, mechanical literacy may be disappearing from the minds of many. Such declines in traditional mechanical complexity, however, might be more than offset by newer forms of technology-related literacies. It therefore remains unclear whether we should condemn or celebrate these changes. At the least, older adults need to understand the possible roles they can play in supporting the interests and skills of those who follow them.

What Kinds of Explanations and Insights Will Future Generations Want?

The decline in experiences with explainable mechanisms presumably affects views of explanations. Younger generations may come to see explanations as primarily providing insights about functions and user procedures while earlier generations place more emphasis on mechanisms. Imagine seeing a robotic lawn mower for the first time. Today's young adults are more inclined to seek explanations of how to use the mower and care less about how the mower works. In contrast, people from older generations are more likely to also want to know something about what enables the lawn mower to perform its functions. For example, they might wonder about the drive mechanism and how easy it is to repair. Perhaps areas of expertise would branch off in different directions, with young adults learning about more subtle functional wrinkles such as using the mower to create attractive patterns in the lawn, while older adults would learn more about how to repair them. Younger adults might have less well-developed abilities to perceive the affordances of parts and subassemblies because they have rarely seen ways that power is amplified, speed is regulated, or rotation is enabled by axles. We will see in the next chapter some signs of such generational differences when we consider internet search query results favored by different generations.

One possible consequence of shrinking knowledge of models and mechanisms in the greater population is a shift in how explanations are evaluated. We saw in chapter 2 that evaluations of explanations can be sorted into those that consider beauty and those that evaluate fits. In addition, explanations are subject to overall interpretative biases and strategies that may negatively influence evaluations. In all these ways, differences between generations are suggested.

Beauty Across Generations

Coherence is central to high quality explanations. The elements in an explanation should fit together to create a larger structure in which elements are not in tension with each other. We prefer explanations in which statements provide mutual support for an overall argument rather than just appear as a list.[21] For example, suppose you wanted an explanation of how a grandfather clock worked and were told statements about the pendulum, the escapement, the hour and minute hands, and the springs, but were given no information on how they are related. That response would be less coherent than one describing how the components connected to each other. Specific connections yield a high-level framework with temporal and spatial sequences. That framework in turn provides insight into what is most important to learn. Coherence can be apparent even when there are major gaps in an explanation. If elements are embedded in an overarching narrative that makes sense, the gaps may be treated as variables covered by a functional gloss ("it somehow regulates speed") to be further unpacked mechanistically.

How then might generations differ in coherence evaluations? If the most recent generations have experienced a drop in mechanistic explanatory understanding, coherence might be judged solely in terms of functional coherence or in terms of the shallowest aspects of mechanisms. Explanations that are appealing to those with minimal mechanistic knowledge might be seen as poor by those with greater mechanistic knowledge. Conversely, those with minimal mechanistic knowledge might reject explanations that seem to have disparate elements when considered at the surface level, but which are compelling at deeper levels to those with greater mechanistic knowledge. This may favor mechanistically knowledgeable people in many cases involving induction, but in other cases, the appeals of deeper similarities might be misleading and put such people at a disadvantage, perhaps leading them to erroneously infer elegant efficient mechanisms in some biological systems.[22]

Simplicity and complexity evaluations would probably vary across generations like coherence evaluations. Thus a person with weak mechanical knowledge might regard an explanation of seemingly diverse surface parts and intricate internal mechanisms as complex while another person with extensive mechanical knowledge might see how the internal mechanisms are all variants of one simple causal system. Context also matters; simply providing a mechanistic explanation can shift preferences from simplicity to complexity.[23]

Finally, breadth impressions could vary systematically with degree of mechanistic knowledge. Consider a case where an explanation of how an industrial silo works refers to an auger that lifts the grain up from the silo to a horizontal conveyor. If the listener had no sense of how augers belong to the classical machine group known as screws and didn't know how screws work, they might not realize how broadly the basic transport mechanism generalizes and think it works only for grain elevators. Screw conveyors are remarkably general-purpose and are used for an enormous range of substances, such as gravel, powders, and fluids.[24]

Fits and Misfits Across Generations

People with richer mechanistic understandings of how well explanations fit with real-world categories may differ considerably from novices. However, while generally more helpful, greater understandings may also handicap a knowledgeable person. Those with more elaborate understandings of different kinds of mechanical systems may over-apply mechanical explanations to biological processes. For example, expert chemists' greater facility with protein folding models may enable them to postulate clever mechanisms of muscle contraction that overlook the more haphazard ways that such systems have emerged through evolution.[25]

Interpretative Biases and Strategies

To resist biases like essentialism and reductionism, alternative explanations must be easily available. Traditionally, as children gained greater causal knowledge about an area, they were more able to reflect on the relative feasibility of different accounts. Since much of that knowledge has typically included mechanistic explanations, if a later generation had weaker mechanistic understandings, its members might be more attracted to explanations reflecting early-emerging biases. They could possibly offset the declining explanatory knowledge with some other forms of elaborated knowledge or refined skills, but there is at present no consensus on what those other forms might be.

For the past five years, I have taught a seminar for college seniors in which I have asked the students to describe how well they understand the mechanisms underlying natural and human-made things. Many happily admit to having minimal knowledge and little interest in how modern devices work. They justify this view on two grounds related to strategy. First, as science and technology have evolved to ever greater levels of complexity, one must find shortcuts around the ever more challenging and time intensive activities required to maintain literacy in all these areas. Second, good causal explanations are available online, and AI systems can now rephrase explanations adjusted to whatever level a novice needs to gain an approximate understanding. It is therefore pointless, and indeed wasteful, to spend time trying to acquire and maintain the full spectrum of such knowledge in one's mind. This is an interesting claim. An increased outsourcing of mechanistic understanding might be an adaptive and efficient way of coping with the complexities of today's sciences and technologies. But costs would seem to be incurred when explanatory understanding is a few keystrokes or swipes away instead of residing directly in one's own mind.

Across several different factors that influence our judgments of the quality of explanations, generational differences in expertise evaluations are consistently plausible. Many of these relate to shifts in emphasis from mechanistic to functional explanations. To consider possible implications for daily life, we turn to an area of growing concern: misinformation.

Misinformation and Mechanisms

An explosion of recent research on the effects of misinformation has occurred not only across many areas of psychology, but also in fields such as political science, economics, climate change, and medicine.[26] Much of the surge in studies is a response to the proliferation of false claims that took off during the COVID-19 pandemic. Whether described as misinformation, disinformation, or fake news, falsehoods can rapidly become entrenched in large sectors of the public. Attempts to inoculate people against misinformation often focus on the social and emotional levers of motivated reasoning; other approaches stress evaluating the coherence of the message.[27] Coherence may be where mechanistic insight is especially useful. Underlying mechanisms provide a deeper level of coherence that is more difficult for purveyors of misinformation to create and effectively disseminate.

It is easier to produce a functionally coherent set of misleading claims when those claims do not also have to be mechanistically coherent. For example, imagine a claim that COVID vaccines cause autism that is supported by claims about how vaccines function by identifying unusual patterns in the body and destroying them. As a result, people with slightly different brain circuits will have those circuits damaged much more extensively than those with standard circuits. This false account may seem functionally plausible but quickly breaks down with a little knowledge of how vaccines for viruses work.

Studies on the effectiveness of correction attempts find that corrections to misinformation work best when they provide an alternative explanation to a false claim.[28] Suppose cell phones are claimed to be harmful because a nefarious group is using the phone's electromagnetic transmissions to compromise the health of certain sectors of the population. A correction without an explanation might simply say that the claim has repeatedly been shown to be false. An explanation might provide alternative reasons why heavy cell phone users might seem to have greater health problems, such as showing that they are less likely to engage in vigorous exercise. For such explanations to work, they will need to show how less exercise is more likely to cause health problems than the low levels of electromagnetic radiation emitted by cell phones.

Inoculating people by warning them about in-group partisan effects and emotional arousal can be difficult to calibrate in daily life. When confronted with a message from a group that has been negatively portrayed as dangerous and deceptive, strong skepticism may be appropriate; but it may also lead to neglect of our own biases. In contrast, coherence may be more consistently invoked as a lens for evaluating all claims regardless of the source. Assessing coherence not only helps vet a message; it also reveals avenues of further exploration. For example, if a claim is rapidly spreading about the dangers of a vaccine, both supporters and skeptics should want to understand the proposed mechanism and its connection to evidence.

In the case of COVID-19 vaccines, false statements about the likelihood of dangerous side effects almost never supported those claims with descriptions of plausible mechanisms responsible for the side effects. Moreover, implausible mechanisms were rarely justified with supporting objective evidence. Underlying mechanisms were also missing for claims that the vaccines contained tiny tracking devices. Similarly, claims spread early in the pandemic that wearing a mask "activated" the virus even though no plausible underlying mechanisms were supplied that explained how this activation could occur.[29]

Everyone should be interested in acquiring some sense of underlying mechanisms regardless of their political leanings and their social biases. Deeper understanding benefits all members of society. But perhaps an emphasis on mechanism asks too much of nonexperts. Many scientists might consider it folly for most of the public to be able to understand, let alone evaluate, mechanisms. This is a critical issue across the sciences. No one expects novices to have to follow complex equations, or four-way causal interactions, or complicated models that confound experts as well; much of the time, brief causal sketches and simplified idealized models may be enough for people to participate in discussions.

Consider, for example, the claim that face masks activated the COVID-19 virus. How could putting a mask over the nose and mouth activate a previously dormant virus? Some viruses can become dormant and be reactivated, but that is almost always achieved by specific triggers such as the presence of certain proteins or a change in ambient temperature. It is difficult to see how face masks could be uniquely responsible for the increased production of certain proteins. Similarly, if warming the upper respiratory tract increased risk, why wouldn't the virus be triggered by scarves, hot drinks, warm showers, and many other agents that could cause comparable local warming? A rough sense of how viruses work and an awareness of well-known dormancy cases, such as occurs with HIV, would challenge the mask claim and would enable focused questions on how that account could possibly be true.

Highly implausible mechanisms sometimes turn out to be true, but high implausibility warrants increased skepticism and seeking more details. Most people may not naturally seek further details or try to find evidence for intuitively implausible mechanisms, but with a little practice, doing so can become more natural as well as rewarding since they expand mechanistic understanding.[30] Inoculating people against misinformation is not easy. There are well-known backfire effects where interventions make things worse.[31] Thus, when parents were given information refuting a link between vaccines and autism, some parents who were initially strongly opposed to vaccines became even less inclined to seek vaccinations for their children.[32] Such effects are uncommon, however, in comparison to the positive effects of refuting misinformation.[33] In general, more reflective cognition and domain-relevant knowledge can reduce negative effects. All this may be related to intellectual humility and having a positive attitude toward one's own ignorance.

As individuals, people can try to defend against misinformation by engaging in *lateral reading*, in which they attempt to systematically search through a wide range of sources. Doing so, however, may require considerable prior knowledge. Suppose I read an article claiming that vaccines cause autism at alarming rates and I know that the claim is controversial. How do I know what other sources to read? Should I look across the age range of authors, authors' religions, authors' home countries, or perhaps media formats, such as videos and text? While each of these ranges may reveal unique arguments, some kinds of ranges will be more informative than others. In this case, we might expect a wide survey of sources from different countries would provide the most informative spread of arguments. Because of the challenges of knowing the right range of sources, we may need to use additional heuristics to find a good spectrum. Looking for different classes of mechanisms might be one such strategy.

One might try to address misinformation by consulting only high-status, peer-reviewed sources. This seems like a reasonable heuristic, but again it can be difficult for laypeople to follow.[34] Knowing good sources may again rely on a sense of the key mechanisms in a science. Most college graduates cannot distinguish high-quality journals from poor-quality ones, including journals in their own major.

Analogical Launchpads

Analogies are central to creativity and innovation. Their power arises by highlighting a critical relation that, when considered at a sufficiently abstract level, allows people to see how the relation bears on a new domain and thereby provides insight. For that reason, analogies frequently support the teaching of causal relations. The more one knows about underlying causal relations, the more one can use this knowledge to gain insights into new domains involving similar causal relations. Relations are far more than simple facts. They make analogies work.[35] To better see how generational effects related to analogies might occur, we need a closer look at analogy generation and use.

Over a hundred years ago, the physicist Ernest Rutherford described the structure of the atom with an analogy to the structure of the solar system. The sun corresponded to the atomic nucleus and the planets to the electrons orbiting around the nucleus. As for all analogies, only some aspects of the original situation applied to the new one. Thus, atomic nuclei are not meant to be hot or yellow. In addition, many features of atoms are not covered, such as the different components of the nucleus and the charges of electrons and protons. Transfer between the two domains is not automatic and can lead to distortions, such as the false inference that gravitational forces explain the interactions between parts of atoms. Distortions also occur when the elements in relations are assumed to share more properties than they do. While most people seem to intuitively realize that atomic nuclei are neither hot nor yellow, they do often assume that electrons are bounded objects, like planets, when in fact they are more like probabilistic clouds representing where electrons are likely to be found. Electrons do not orbit the nucleus along specific paths (as planets orbit the sun); this is where the solar system analogy breaks down and misleads.

Analogies are essential to creativity, innovation, and instruction. Their power arises from highlighting critical relations. When a relation is grasped at a sufficiently abstract level, it provides insight into how the relation holds in a new domain. While analogies can be powerful vehicles for providing new insights in instruction, most teachers have difficulty generating new analogies. Thus it is easy to tell students about the solar system as a way of describing some features of atoms. It is much harder for an instructor to think of a new analogy to explain how a GPS device works or how a rainbow forms. (Describing water particles as tiny prisms is not an analogy. It is a restatement of the effect.) One way to facilitate the creation of new analogies involves developing detailed understandings of the mechanisms underlying systems across diverse categories. Teachers who model such broad and deep backgrounds can be inspirational.

Analogies between two domains are often more useful in one direction than the other. Thus, while many mechanical analogies give insight into psychological and biological systems, psychological analogies may be less likely to illuminate mechanical systems. The movements of pendulums might serve as an analogy for understanding mood swings; using mood swings as an analogical base for understanding pendulums seems odd. These apparent asymmetries reinforce the proposal that physical mechanical understandings may be especially potent sources for analogies to provide insights into distant domains.

Asymmetries in the informativeness of analogies are usually explained by the idea that the "base" domain is more feature rich, coherent, or systematic than the target domain. If there are consistent asymmetries between one domain and several others, those asymmetries illustrate how that domain has especially well-structured relations for serving as conceptual launchpads for analogies. A precedent for such asymmetries is with similarity ratings. The psychologist Amos Tversky showed this with examples such as North Korea being seen as more similar to China than China is to North Korea. Similarly, if one item has more elaborated relational knowledge than the other, it should serve as a more effective base for learning about the other. The psychologist Dedre Gentner has noted this effect in terms of informativity and has shown that people consistently prefer one direction over the other for many specific analogies.[36] But what about entire domains? If humans are especially gifted at acquiring relational causal models in the realm of physical mechanics, they should be more prone to using such models to gain insight into other domains than vice versa. A full set of formal experimental studies is still needed to support this prediction, but readers may find some more informal arguments suggestive.

We can use large language models that have proliferated in AI systems to look for domain asymmetries in analogical preferences and informativeness. For example, we can ask ChatGPT-5 to generate ten analogies explaining psychological systems in terms of physical devices in the following way: "Provide ten different analogies that explain psychological phenomena in terms of physical mechanical models." In an informal survey, most of the analogies provided by GPT-5, over several trials, were rated by adults as somewhat useful or even insightful (e.g., explaining attention in physical terms: "A focusing lens or movable spotlight can narrow or widen its beam to illuminate a chosen area. Similarly, attention can concentrate narrowly on a single thought or widen to scan broadly. Both involve selective amplification of a limited region.").

Now consider what happens when we ask GPT-5 to generate analogies explaining physical systems in terms of psychological ones: "Provide ten different analogies that explain physical mechanical phenomena in terms of psychological models." Most of the analogies generated are viewed by adults as less useful or simply misguided (e.g., explaining friction in psychological terms: "When two surfaces rub, motion is slowed by friction—just as new ideas or behaviors encounter social or mental resistance. Friction and conservatism both dissipate energy as heat."). Because GPT-5 and similar AI programs are based on naturally occurring text strings generated by humans, they provide a window into what domains are especially powerful sources of analogies for understanding other domains. We are more prone to make mechanical analogies about psychological phenomena than the opposite, presumably because we have an especially well-articulated set of mechanical intuitions. The asymmetry may start to diminish over time as mechanical knowledge shrinks. If humans naturally rely on early understandings of physical relations to understand aspects of psychological relations, and if they do this more fluently than the reverse, people with weaker physical mechanical models may gain fewer insights into the nonphysical.[37] The asymmetry will not hold in all instances, and physical mechanics is not likely to be the only fertile launchpad for analogical insights, but the pattern holds for a wide range of domain pairs.

While clockwork mechanisms may provide useful analogical insights into psychology, they are not sufficient for grounding all intuitions about the mind and behavior. When considered in more detail, parts of folk psychology do not fit neatly with layered hierarchical mechanistic models. Instead, accounts often rely on narrative structures that can also serve as platforms for new insights. A recent resurgence of interest in the role of narrative structures suggests that they may be uniquely suited for reasoning about multifactor causal systems where tabulations of multiple probabilities are cognitively overwhelming.[38] Narratives are rarely simple linear causal chains and instead tend to have elaborate interwoven sets of multiple causes and consequences. As we saw earlier, narratives may be useful in understanding aspects of the human immune system.

Almost all studies to date on the cognitive effects of new technologies have asked how device changes might influence general skills in areas such as attention, memory, and reasoning. They rarely explore whether there might be consequences for broad classes of causal models that support analogies. One reason for this neglect may be a focus on specific areas of expert skills. If studies were to examine effects relating to declines in activities such as punch card and VCR programming, those studies would uncover little of interest. Skills in such local domains will obviously decline with little or no consequence for other cognitive abilities. Our focus, however, is on a broader type of domain, namely causal mechanistic understanding. Mechanistic thought involves formats similar to clockwork mechanisms, and the digital revolution has resulted in a widespread loss of experiences with such mechanisms. From toasters to alarm clocks to cars, zoomers are less prone to encounter or talk about underlying mechanisms, including those that have stayed essentially the same for the past fifty years, such as automobile suspensions.

Declines in understanding at this broader mechanistic level arise from different experiences across a wide range of devices. As such, they should result in a diminished facility with certain causal relational schema that frequently recur across devices. Examples include ratchets, screws, levers, gears, pulleys, and reservoirs. This broad domain has become one that humans are especially adept at learning about just because artifacts were designed and built by humans embedded in societies. The domain has been sculpted by the more conceptually fluent ways that humans process and represent causal patterns and the ways that cultures can best amplify and transmit that information. Several researchers from psychology and computer science have suggested that a key part of this process is a very early emerging, and perhaps innate, core understanding of aspects of physical mechanics. But even if some of this knowledge is innate, it still requires sufficient relevant experiences to develop, just as neonatal perceptual abilities can wither in impoverished environments.

If zoomers do have diminished understandings of underlying physical mechanisms, that deficit may shape their ability to learn from and communicate with analogies. Such effects would occur if knowledge of physical mechanical relations play an outsized role in providing the most fertile and insightful analogies. This possibility has not yet been subject to much experimental study, but current AI models that rely on massive samples of human language use suggest that analogies drawing on physical mechanical relations tend to be seen as especially informative.

* * *

Missing mechanisms deprive young adults of a powerful way of combating misinformation. We don't need to know endless details about specific classes of mechanisms—just enough to assess the plausibility of claims about important issues such as vaccines. Cumulative cultural evolution in humans may have taken off not just because of increased theory of mind skills but also because of enhanced technology skills.[39] Progress, however, is not inevitable, as in cases such as the decline in prehistoric Tasmania discussed earlier.

Effects on creativity and innovation might follow from a shifting base of fluent relational models. Analogical thought can inspire new ideas and provide novel solutions to problems. Mechanistic knowledge has been an especially potent source of insightful analogies; it seems to afford more useful analogies into other domains than other domains do into mechanistic understanding.

Our discussion has explored how new technologies might transform the activities of individual minds. Yet the most significant consequences of technology occur when groups interact and people come together to solve problems. Blanket statements that technology is all good or bad have not fared well historically. The key issues almost invariably concern the trade-offs involved. Those trade-offs may loom especially large when technology influences how minds interact in groups during young adulthood and beyond. We turn to those processes in chapter 8.

8 Interpersonal and Outsourced Thought Across Generations

Mobile phone use exploded at the dawn of the twenty-first century. Most people alive today can hardly imagine a world without such phones. Conversely, few people in 1980 could have imagined that by 2025, more than 95 percent of the world's population would either have their own mobile phones or have easy access to them. Today, most of those phones are smartphones, devices with ever multiplying capacities ranging from star finding to movie making. The cell phone revolution illustrates how generational cognitive differences resulting from new technologies can be amplified when they are embedded in interpersonal interactions, not just in the minds of isolated individuals.

Rules of Digital Civility

One cognitive consequence of growing up with cell phones involves changes in how people arrange meetings. In 1990, if you learned that you and a friend would be visiting New York City at the same time, you might arrange a meeting in several steps. You might first agree on a meeting point, such as a large museum that both of you knew. You might then agree on where in the museum you would meet. Finally, you would decide on a meeting time and perhaps on how long either party would wait before giving up; and you might set up a backup option if one party failed to show.

Arranging this meeting draws on several aspects of cognition, including cognitive maps, knowledge of popular locations, and a sense of meeting etiquette, among many others. Such meetings also connect to a classic paper by the philosopher David Lewis on social conventions.[1] Lewis described situations where people are confronted with *coordination problems*. They solve these problems by relying on common knowledge of social conventions.

When arranging a meeting, both parties must sense what location would be reasonable not just for themselves but also for the other person. They should also sense what both of them know about—for example, the internal layout of the museum, what exhibits are both popular and permanent, and how crowded certain exhibits might be at various times. Both parties should also pick an appropriately sized location with clear boundaries that make finding the other person easy. A location such as all of Central Park would be too large. Times Square is a more feasible size, but it often has too many people present to easily pick out a friend. Hundreds of more subtle shared assumptions are also at work, such as the physical and economic challenges of getting to a location and whether it will be safe for both parties.

In 2025, when two friends want to meet up in New York, the advance arrangements might simply be to agree to text or call on the day that both will be in the city and then work out a good place and time to meet, often following that initial contact with brief calls and texts. I first became aware of this new way of getting together in 1999. I had just purchased my first cell phone and was trying to arrange a meeting with my oldest son, Derek, who was a sophomore in college and had been using a cell phone for about a year. We discussed our plans in person, but when I started talking about possible locations and times, Derek quickly interrupted. He told me to just have my cell phone on during that day and we'd zero in on each other.

Derek's meetup strategy is the new normal among most adults. It comes with the obvious advantages of being able to quickly adjust meeting times and places in response to new developments such as traffic jams or weather changes. Several studies have documented comparable strategies in many naturalistic settings.[2] At first glance, this looks like a clear advance thanks to the new technology; but for all its appeals, meeting and traveling with less advance planning can have cognitive costs. For example, habitual use of GPS apps on phones negatively affects spatial memory.[3] One of our other sons, Dylan, freely admits that he has almost no cognitive maps of anywhere, including his own neighborhood, because he always follows instructions issued by his GPS in real time. That said, he is a gifted GPS navigator, adapting instantly to rapidly changing situations.

The switch in meeting arrangements might result in less conventional knowledge of the spatial layouts of locations within larger cities and neighborhoods. Such spatial knowledge does seem to be degraded in zoomers, but the losses may be offset by newer convention-laden skills. For example, new subtle forms of convention tracking have appeared with respect to the kinds of messages to use in communications.[4] Zoomers have better developed senses of digital communication etiquette. A 2024 *Wall Street Journal* commentary illustrated this trend with the title "Don't You Dare Call Me Without Texting First."[5] The earliest articles about texting first appeared around 2015.

New systems of etiquette have emerged as communication channels have diversified. Communication can occur through letters, email, texts, social media, enterprise versions of social media (e.g., Slack), video games, one-to-many broadcasts (e.g., tweets), many-to-one messaging (e.g., bullying), live video on phones, and live video on the internet (e.g., Zoom). Some are *synchronous*, meaning real-time back-and-forth interactions. Others have intrinsic delays and are *asynchronous*. Some promise great privacy, others advertise great notoriety. The constantly evolving rules of etiquette are more actively tracked by younger generations. As a boomer whose profession requires frequent communication with zoomers, I'm often aware that I'm not in synch with current norms. Zoomers almost never contact peers by phone without texting first; even worse is contacting someone first by live video. Email is regarded as formal and stilted, and many undergraduates will confess to not having checked their email for days.

Rules have emerged around acceptable response latencies. It is apparently rude for millennials not to respond to texts almost immediately in peer-to-peer messaging; with older peers, however, and with texts between millennials and those of older ages, longer delays are tolerated. Some zoomers apparently think millennials are too neurotic about responding to texts quickly; other sources regard zoomers as also excessively bound to quick text responses.[6] It is unclear whether zoomers have norms on when to communicate by physical letters and when to use handwritten notes versus typed versions—variations important to the silent generation and boomers.

Some new kinds of knowledge and cognitive skills may be present in generations that employ more complex and multilayered levels of digital etiquette that must also be frequently updated. Fans of the writer Amos Towles and his books *Rules of Civility* and *A Gentleman from Moscow* might applaud these more subtle and elaborate codes of conduct as restoring dignity to the social milieu. Yet being more intuitively socially graceful doesn't come to mind as a hallmark of zoomers. Of course, older generations simply may not be able to see all that zoomers do. When I ask college seniors what they know more about than older generations, they often mention having a better grasp of social nuances.

In terms of costs, one deficit could be the increased cognitive load imposed by tracking all these communication guidelines in the moment. Those loads might result in less cognitive resources being available for other activities such as engaging in deeper reflection and unpacking subtle causes. Perhaps zoomers take more care to avoid embarrassing themselves and others, thereby focusing more on social and psychological relations and less on physical context.

Etiquette shifts have followed technology changes in the past. When we moved to Ithaca, New York, in 1977, our first phone service was a party line. Apparently, the prior resident at our address liked the cost savings. Party line members shared the same phone number with other homes and were expected to follow certain protocols such as hanging up on a call quickly if you picked up first and it wasn't for anyone in your home. You then did not answer when it rang again right after, with the longer sequences of rings indicating to other homes that it was for them.

When people shifted to one number per house, there were extra charges for additional phones or "extensions" beyond the main house phone. Those who lived for several decades with one phone for an entire household learned, when making outgoing calls to similar households, to prepare for possible conversations with all the members of the other household who might answer. In my first year in college, I lived in a fifty-student community with one phone. Each of us had phone duty for a few hours each week, during which we would sit by the phone and run to get the person being called. Pretty much the entire community had a constantly updated knowledge of who was receiving and making calls and who the likely call partners were. It sounds intrusive today, but it also bound us together as we developed a sense of each community member's friends and families.

Telephones were hardly the first innovation in communication that resulted in behavioral and cognitive changes. Before phones, some people developed ways of notifying others that more intimate communications were forthcoming. For example, since at least the fifteenth century in China, potential visitors engaged in a series of precursor interactions through "calling cards," with quite complex details unique to each culture. In Victorian England, bending a corner of the card indicated that you were waiting in person while a servant handed the card to someone in the home.

Yet the scope and breadth of change in communication practices in the twenty-first century seem larger, and they suggest substantial shifts in how people think when in the company of others.

While our focus here is on teens and adults, the first awareness of etiquette and social conventions appear earlier. Young children have reliable intuitions on how to solve coordination problems. In a pioneering study building off Lewis's philosophical works, Kristi Lockhart led a team showing that, between first and third grade, children in the United States improve greatly in the ability to distinguish social conventions from physical regularities and moral laws.[7] For example, children were asked to suppose that everyone got together and agreed to drive on the left side of the road instead of the right side. They were then asked if people could make that change. They were also asked to imagine that everyone got together and agreed that rocks could float, or that it was okay to steal from others, and were then asked if people could make that change. Between the first and the third grades, children learn how to keep social conventions distinct from the other two kinds of regularities.

Lockhart went on to conduct studies showing that children as young as five sense that the ways one explains violations of social conventions are distinct from the ways one explains violations of physical regularities and moral laws.[8] The same children also had a rudimentary sense of how to solve coordination problems, which abound in social interactions. For example, if a child learned that another child had a strong preference for the color red and they are put in a game where both children would win if they picked chips of the same color, the child would choose red chips. These studies, and many others that followed, show an early sensitivity to different kinds of causal patterns relating to social situations.[9] They show that components necessary for social rituals are present in early childhood and develop into ever more refined forms for several years afterwards.

Transactive Thought

The psychologist Daniel Wegner used the term *transactive memory* to describe how a group of people cooperate to reduce the memory load on individuals in the group.[10] Transactive memory has three components: the memories of each member of a group, the ways that memories across individuals interact, and the ways that people keep track of what each person knows even as it changes. Transactive memory works best when members of a group know each other well and can more efficiently divide up memory duties. Thus, longtime partners do better than strangers or casual acquaintances in recalling new information that they have encountered together.[11] We have all seen how couples that have been together for decades can seamlessly complement each other's memories in real time in a conversation, automatically filling gaps in what their partner has said.

It is useful here to introduce a wider sense of *transactive thought* as a collective cognitive system where individuals within a group rely on one another to process, store, and retrieve information. This system goes beyond mere memory storage; it involves the dynamic interplay of cognitive processes across multiple minds, allowing for shared problem-solving, decision-making, and knowledge creation. Each member contributes their cognitive strengths, creating a network where the group's collective intelligence surpasses that of any individual. Through communication and collaboration, transactive cognition facilitates efficient handling of complex tasks and enhances the overall cognitive capacity of the group. While transactive memory typically refers to groups with many hours of shared experience, transactive thought applies to any case where groups outperform individuals. Thus even strangers may outperform themselves as individuals by tapping into shared conventions on how to distribute cognitive labor.[12]

When our children were young, my wife (Kristi Lockhart) and I learned on short notice that our day care center was being evicted and that the only available and affordable location was the county jail. The jail had recently been vacated after the county decided it was below minimal standards for housing prisoners and built a new jail at a new location. At a hastily called parents' meeting in the former jail, we were given a few weeks to transform that grim space into something suitable for our children. The twenty or so parents present immediately identified needed skills ranging from carpentry to metalworking, drywall installation, and code compliance checking. By the end of that meeting, we had sorted ourselves into expertise clusters. After additional expertise tuning, we met the deadline with a quirky yet oddly welcoming space in which even the remaining cell bars served as colorful architectural features.

Transactive thought occurs whenever groups develop patterns of interaction that take advantage of the different cognitive assets of members. Those interactions can develop implicitly. Even when those patterns are sophisticated and rapid, they may have occurred outside of awareness. Other times, explicit negotiations may be critical to distributions of cognitive activities and the knowledge endowments of each member.

Many kinds of stored knowledge contribute to transactive thought, not just memories of facts or events. It can rely on the understandings and insights unique to each member of a group. Because it can amplify the power of thought, it can enhance understandings and insights in ways that go beyond what is possible for individuals. Yet people tend to underestimate the extent to which their cognitive abilities and achievements depend on other minds; they are frequently surprised at how much their performance suffers when they are on their own.[13]

Dividing and Sharing Cognitive Labor

We saw in earlier chapters how divisions of physical labor have organized societies throughout history. As groups of humans came together in larger, more stable communities, they benefited when different members of those communities became experts in specific skills, such as tending fire, tracking animals, and constructing dwellings. These divisions of physical labor required accompanying divisions of cognitive labor. In the early periods of human existence, this expertise may have been mostly procedural.[14] Experts learned how to perform action sequences with precision and consistency. They usually started learning those skills as children, often as apprentices.

Adults may scaffold skill learning by ensuring that the required materials and components are easily available at the right times, but scaffolding did not require explicit descriptions. Over millennia, expertise became more heavily infused with explicit or declarative components. Experts started to explain verbally how and why they did things. Declarative expertise enabled accelerated rates of learning and greater preservation and transmission of knowledge across generations. This happened first in oral forms in preliterate societies and then in text, even more effectively, when writing systems emerged.

The advantages conferred by divisions of labor have been described by Adam Smith and his successors as greater productivity, economy, and quality.[15] But those salient advantages can overshadow an important component—the need for a common cognitive ground of shared knowledge that integrates experts and connects them to laypeople. If a group of IT experts wants to persuade a town to upgrade its police department's laptop computers, the experts need not only their shared knowledge about laptops but also to know what aspects of that shared knowledge the town's citizens have.

Most nonexperts today have considerable shared background knowledge of laptops' functions and operating constraints. For example, they understand the roles of writing and simple mathematics. They know that electricity powers most devices and that, if they are portable, those devices have batteries. Most people have some sense of why there must be capacity limitations, the pluses and minuses of cloud storage, why laptops need to be periodically charged, why they usually aren't waterproof, and why dropping them on hard surfaces may cause irreparable damage. These basic intuitions sit on top of more detailed knowledge about features of certain laptops that experts would want to convey. In contrast, a farmer living in France in 1250 would need massive amounts of new knowledge to make any sense of a modern laptop purchase. We continuously calibrate what we say to others through inferences about their relative level of expertise.

More broadly, a community should know when it is appropriate for there to be just a few superbly trained experts in an area with everyone else knowing almost nothing, and when the entire community should know some basic principles and causal patterns. We should also have a sense of when experts might be expected to know just enough to meet practical demands and when they should have additional knowledge that will support continuing innovation. Sometimes we need several levels of expertise, which in turn requires knowing the best ways to organize them.

To get a better sense of these kinds of issues, consider the case of expertise in medicine. In more developed societies, the degree of expert specialization has accelerated over the past few decades.[16] In 1960, most ophthalmologists were expected to be experts on all aspects of eye care, ranging from blocked tear drainage to scratched corneas, cataracts, glaucoma, and detached retinas. Today there are specialists for each of those eye problems (and others) who do nothing but treat those specific problems. This hyper-specialization has led to a proliferation of ever narrower journals and professional groups and to specialists becoming increasingly ignorant of other eye problems and their treatments. This can result in a kind of tunnel vision or myopia concerning what eye problems are most important. A different kind of eye expert has also coevolved: people who know as much as possible about the full spectrum of eye diseases. Today those experts are often senior ophthalmologists from prior generations, but they are rapidly disappearing, even though their form of expertise might be critical for cases that straddle boundaries or are completely novel.

The general practitioner or family medicine physician represents the broad expertise that most people see as essential to medicine today. A middle tier of relatively broad specialists in areas such as vision or general surgery may also become more important. At the least, broader expertise might be distributed transactively by having eye experts from different specialties meet to review cases together. The need for regular meetings of such teams is starting to be recognized in medicine.[17] These kinds of expertise clusters may emerge organically in some professions, but they usually occur haphazardly. Given wide agreement that well-functioning divisions of cognitive labor may be the most critical component for the continued development of cultures, surprisingly little attention has been paid to the optimal structuring of expert clusters. The rate of innovation may depend on how many experts there are at different levels of specialization. As greater densities of experts in communities are related to higher rates of innovations, we would like to know how to best distribute those clusters.[18]

Medicine is hardly unique in having to face the challenges of determining optimal levels of expertise and how they should be populated. Comparable situations can be found in computer programming, architecture, and law, to list just a few.[19] In all cases, other influences, such as financial rewards, can affect the distributions of specialists in ways that work against the best arrangement.

Different amounts of mechanistic understanding could have profound effects on how expertise clusters as children transition to their adult lives. If most members of a population have minimal understanding of how things work, they might default to thinking of expertise in more functional terms. When trying to decide to invest in a new technology, such as the latest generation of laptops, they might focus on performance metrics such as speed and storage capacity for applications that they will use. Such strategies can be highly effective, but in some cases, understanding how the device works may be critical. If you understand how a new kind of graphics chip enables more parallel processing and you understand how that processing will be central to a new wave of programs, your laptop purchases may be far more effective than those of someone who has no knowledge along those lines. Similarly, if you are deciding in purely functional terms whether to get a vaccine booster, you might simply look at the reduced likelihood of catching a particular virus. But if you know something about how the vaccine works and the biology of the virus, you might see how the mutation rate and vaccine revision schedules suggest a better strategy.

Conceptual Common Ground

An essential part of transactive thought is establishing an appropriate common ground of understanding. When the problems confronted by a group are well defined and time-sensitive, a strong level of shared knowledge can improve problem-solving.[20] But having an appropriate level of common knowledge requires a sense of the minimum knowledge that all members of a group should have and what is better left to specialists.

Just a few years ago, trying to find appropriate experts required a shared sense of broad expertise domains and of some subdomains as well. Today, young adults may be more willing to delegate virtually all expertise tracking to artificial intelligences. AI-guided search may be nudging people away from having detailed common ground in their minds and toward greater ignorance. If a problem arises, they may simply describe the problem to an AI interface and ask whom to approach for expert advice. For example, if the heating system in a house stopped working, zoomers might describe the lack of heat to an AI agent and contact its recommended experts. In contrast, older generations would be more likely to first try to zero in on the problem themselves based on basic knowledge of home heating systems. Thus they might check if the problem was due to an empty fuel tank, a malfunctioning thermostat, or a tripped circuit breaker. Those with a bit more knowledge could check for additional possibilities, such as a clogged fuel line, combustion failure, or a faulty emergency switch. Today, such attempts to narrow down the possible causes may be seen as a waste of time when a heating service company can handle all these problems. But a rough understanding of the operations of heating systems can reduce expenses and speed up the repair process. That understanding might also lead to the use of experts who are more relevant to the specific problem, such as electricians.

If zoomers have weaker common understandings of mechanistic patterns, reducing the knowledge requirement might enable them to more quickly establish ad hoc common grounds as the situation demands. Perhaps these adaptive, dynamic common grounds are more responsive to today's rapidly shifting social contexts. Zoomers might also trade weaker shared mechanistic knowledge for stronger shared functional and procedural knowledge. Thus they might know less about how a new microwave works, but more about how to adjust warming options.

As communities grow and develop distinct areas of expertise, interactions are more efficient when experts can rely on concurrent expansions of relevant knowledge in nonexperts. Certain core forms of causal understanding may be especially useful across most areas of expertise and may be repeatedly used to communicate key ideas and serve as sources of productive analogies. Historically, causal understandings derived from the revealed inner workings of everyday objects may have helped to establish common ground. We now must ask whether this form of knowledge is still relevant considering emerging technologies and developments in the sciences.

The psychologist Herbert Clark developed the idea of common ground in his many studies on human communication.[21] Common ground may be negotiated and assembled during initial meetings of a group. Some common ground may be relevant only to the immediate situation and must be constantly updated. Often, many elements must be ascertained and negotiated in parallel. Many of these common ground practices are present in young children.[22] A three-year-old may hold up a kind of toy and ask a new acquaintance if they have one like it. If they do, the conversation proceeds differently than if they did not.[23]

Finding linguistic common ground can be as straightforward as determining what language will facilitate the most informative interchanges among all participants. When all members of a group are multilingual, this can be more complex and may shift with the content of the conversation. Many aspects of language are coordinated, including syntactic styles. For given topics, members may favor specific dialects, levels of formality, and technical or nontechnical vocabularies. In addition, members of a group may work to establish shared cultural norms and practices, life experiences, and aspects of the immediate context. Converging on common ground can be essential to successful conversations and problem-solving.[24]

Establishing the appropriate conceptual common ground can be critical to achieving new insights. When talking with experts, laypeople can quickly be overwhelmed if the experts have not adequately ascertained what terms to use and whether everyone has the same meanings for those terms. A novice and an expert chemist may have a very different sense of the word *organic*. Experts may interpret familiar words, such as *digest*, in terms of more elaborate causal underpinnings. In establishing conceptual grounds, participants should have a shared sense of what all members know immediately as opposed to what must be acquired from a nearby resource. Access to knowledge in one's own head may be essential for an efficient discussion. In addition, members of a group may also learn where another member has especially deep conceptual knowledge and naturally shift the conversation to them at appropriate times. These sophisticated dynamics have been documented in studies of groups such as airline flight crews and the bridge staff on large ships. When a group has worked together for long periods, they naturally build a core common ground plus a system of seamless shifting between individuals with more specialized areas of expertise as situations require.[25] Cargo ship bridge teams recognize this need by routinely shifting control from ocean pilots to harbor pilots as they enter a port.

The benefits of common knowledge should not diminish the benefits of group members also having uncommon knowledge. When a group must solve a challenging novel problem, diversity of knowledge may lead to better outcomes. Diversity, however, comes in several forms. People can vary in cultural fluency, disciplinary knowledge, specific cognitive skills, and life experiences. The benefits of both diversity and common knowledge depend on the task and the kinds of diversity and common knowledge needed.[26]

Technical Knowledge as Common Conceptual Ground

In settings involving tools and devices, the cognitive scientist Francois Osiurak and colleagues have explored how a shared background of technical knowledge enhances a group's development of new ideas as well as the transmission of those ideas to others.[27] In one study, small groups called microsocieties were asked to improve a device. They were charged with optimizing the speed at which a multipart wheel rolled down a ramp. Each group member worked for a fixed amount of time on the problem by manipulating the positions of the wheel parts. Their work was then passed to another member of the group, creating a chain of transmission. In one condition, transmission along the chain contained all the information gathered in the prior steps. In the other, "degraded" condition, some information was omitted across transmission steps, as often happens in the real world. Greater technical knowledge predicted better solutions in both conditions, but especially so when information was degraded.[28]

Osiurak's lab has examined transmission chains among members of groups with goals ranging from making paper gliders with the longest flights to building the highest towers out of wires. Greater technical knowledge in all members of a group led to greater rates of knowledge growth and innovation. In one experiment, members of a small group were each evaluated for their degree of technical knowledge, theory of mind skills, and friendliness. The groups then had a sequence of meetings to decide how to build the tallest tower. Over time, the individuals with the most technical knowledge became more influential than those with the most social skills in terms of whose inputs were noticed, remembered, and evident in the result.[29]

If human tool use was accelerated by the emergence through evolution of a uniquely human form of causal mechanical thought, as Osiurak and colleagues have suggested, then having fewer experiences that employ neural circuits specialized for such thought could result in a less-developed brain architecture for supporting mechanical problem-solving.[30] But dedicated circuits are not necessary. A similar effect could occur without brain regions specialized for mechanistic patterns. Long-term experience with any class of information or experiences may affect adult performance for decades; in many cases, deprivation of such experiences from an early age may be especially deleterious. Acquiring aspects of mechanistic thought may be like the ability to learn the accents and syntactic subtleties of languages and dialects, which declines dramatically after puberty. This decline may occur either because certain brain circuits become fixed or simply because early cognitive commitments are increasingly difficult to reverse with age, though they are not strictly irreversible.[31] The enduring importance of early experiences on adult abilities can occur either way.

A widely shared level of technical understanding may have served as a valuable cognitive asset to communities over successive generations. Technical understanding, usually described as knowledge of causal mechanical patterns, augments cumulative cultural evolution and innovation. Its wide scope enables individuals from diverse backgrounds to more easily become involved in new discoveries and inventions. Common technical literacy also makes it easier to record new advances and transmit them to all members of the next generation. But if the opportunities to develop that literacy are disappearing from everyday life, members of current and future generations may have almost complete technical ignorance as individuals. They may instead endorse having many small clusters of people with technical expertise at various levels. They may see it as liberating to no longer need a basic grasp of how things work in the world around them. This perspective, however, assumes the ability to seamlessly outsource understandings to others and to offload cognitive elements to the physical and digital worlds.

Outsourcing and Off-Loading

Cognitive outsourcing typically refers to relying on other minds to provide needed cognitive skills or pieces of knowledge.[32] We engage in cognitive outsourcing when we ask an accountant to do our taxes or ask a cabbie the best route between two points in a city at rush hour. Cognitive off-loading is a related process in which we reduce our cognitive load by relying on support from the physical world.[33] This can be in the form of posting notes on a refrigerator, making a shopping list, or rotating our watch as a reminder to do something. The boundary between outsourcing and off-loading blurs when we carry a phrase book with us in a foreign country or engage in an AI-assisted search.

We can define *transactive thought* as a form of outsourcing that relies on cognitive support from other people in real time. We have seen that it can be especially powerful in adulthood as the parties involved develop, over extended time periods, complementary bodies of knowledge and efficient ways of working together. In a more asynchronous manner, we also rely on products created by others, such as books, blogs, and videos. Outsourcing also has increasingly shifted away from accessing sources that have been edited or curated in some way to requesting information through search engines and social network sites. Most recently, people have been able to access AI resources (a.k.a. chatbots) such as ChatGPT, Claude, Google Gemini, Microsoft Copilot, and Perplexity.

We can off-load directly to physical objects without necessarily involving others, as when we leave notes to ourselves. Since these notes are to our future selves, we can see them as also a way of thinking with others in mind. The potential benefits of off-loading seem clear. By selecting information and depositing it in the world outside our heads, we free up space within our heads both for storage of material and for processing. We may also be providing more durable and accurate records of our knowledge. When we write things down or take photographs, we usually can record information in a more precise and longer-lasting manner than when we try to keep information in our heads.

Cognitive losses can also occur when we outsource and off-load. We may unwittingly shift our focus toward knowing how to access information and away from knowing the actual content. Suppose you have just purchased and moved into an older home. Over the course of a few weeks, you discover many things that need to be fixed or updated. For each of these problems, you write a brief note and deposit it in a safe place. You've learned that you must remember where you put those notes and, as a result, you tend to devalue remembering their content. But this storage strategy can create problems if you're later engaged in a discussion with your partner about renovations and don't have the notes handy, or, even worse, if you deposited the notes in a spot where they got ruined by water damage (one of the things you needed to address) and are no longer legible.

Reliance on external forms of cognitive support can diminish your abilities when you are untethered from that support. Several studies have shown that navigating by following commands barked out by your GPS can result in having more impoverished cognitive maps of your environment.[34] Taking photographs of various objects can result in weaker memories of the objects' properties than if you simply observe the objects for the same amount of time.[35] (The effect can be ameliorated if photographers are told to focus on the parts of objects when taking their pictures. Observers do not need such instructions.) The same pattern extends to attending lectures with large numbers of projected images. Some listeners can be seen taking frantic handwritten notes while others are taking photos of each slide at a more relaxed pace. Notetakers recall the lecture better and have better understandings of the key points.[36]

In terms of generational shifts, compare a teenager taking photographs in 1968 with one taking photographs in 2025. One major difference concerns the number of photos taken. A photographer in 1968, using a 35 mm camera, had to purchase a roll of twenty-four- or thirty-six-shot film. Prints of each photo were expensive, especially if they were in color. I remember often agonizing over whether I could afford the luxury of a thirty-six-exposure roll. Because every photo entailed a significant expense, much thought went into the decision of when to take a picture.

In 2025, a teenager may have never even thought about limits on the number of photos they take. A smartphone can store fifty thousand high-resolution photos and roughly 1,250 minutes of high-quality video. For many teens today, even these are not real limits; cloud services can back up indefinitely large numbers of photos. The only reason for thinking about limiting the number of photos taken might be concerns about finding individual photos in the enormous sea of images. But even that problem is vanishing as AI and other photo search tools can quickly extract an easily searchable set.

These differences must have cognitive consequences. People from different generations have thought differently about the process of taking each shot and what matters most. Some made sure the lighting, pose, and background were the best possible for each photo, while others took as many shots as possible from all angles and settings. Earlier photographers were more intimately related to the mechanisms in the camera and the processes of developing film and making prints. When loading a roll of film into a camera, you could see many of the inner workings of the camera. You had to focus the lens or even swap out different lenses for objects near and far away. For many cameras, you needed to adjust both the speed and the aperture in ways that considered the kind of film and the setting. You knew what the photo lab did, especially if the store where you dropped off the film had questions about the processing. You also had to make choices about the speed of the film depending on the kind of photos you'd be taking (e.g., sunsets versus athletic events). There must be differences in what zoomers and boomers were thinking as they took their photos; but we don't yet know the possible consequences of such differences on memory and attention.

Today, we all have data-rich trails and traces of where we have been and what we have done. Most people have thousands of photos and videos taken by them and others over the years. They have stored somewhere all the texts and emails they have sent and received for much of their lives. We have also left traces we may be unaware of, such as our online shopping and search histories. When asked what we did on a certain date or what we communicated on November 6, 2000, we can search through many kinds of records to find converging evidence. In earlier decades, we might have an old diary or scheduling book to look through and perhaps a few photo albums. If taking photos and auto-logging emails reduces our memory of content, then our mental reliving of the past, and even our dreams, will have fewer details of those experiences when we don't have the photos and emails in front of us. These considerations have led to an area of research examining off-loading and autobiographical memory.

When people want to recall information from their past, their recall is influenced by the degree and kind of off-loading they have done. When they heavily "externalize" their memories through photos, videos, and other resources, their sense of their own past is not the same as of those who rarely, if ever, engage in such practices.[37] Lots of externalizing is happening throughout the world, with many billions of photos and millions of videos uploaded every day. The specific effects vary according to the details of the situation. Taking photos during museum tours can enhance visual memory and diminish auditory memory. In other settings, taking photos can impair visual memory because people are attentionally distracted by the process of taking the photo and devote less attention to encoding the event.[38] Sometimes people deprioritize storage inside their heads when they know they will have an external repository of their experience. In other cases, internal memory loss is primarily driven by attentional limitations. The nature of the photo may also influence memories. If a person has many photos in which they are in the frame either because they took selfies or because someone else took the photo, they are more likely to recall the event from a third-person perspective and may express more positive emotions about the event.[39]

The flood of images and videos of our lives from birth onward may result in future generations having difficulties understanding what has been one of the most robust and fascinating findings in psychology: infantile amnesia. Freud described the phenomenon in 1905 as an inability to remember early experiences.[40] Over the years, researchers have asked people of all ages throughout the world to identify their first memories. Some individuals have false memories of very early memories. People under the influence of LSD and under the guidance of certain psychiatrists have reported recalling such experiences as being pushed down the birth canal and hearing voices in the delivery room.[41] Many studies, however, have shown that when people think backward about their lives, a nearly continuous tapestry of multisensory memories extends back to the later preschool years. But then, around age three, only fragments are recalled, and then nothing earlier. This phenomenon remains a hot research topic in the neurosciences.[42]

That subjective feeling of memories fading away as we think back on our lives may be absent in many of today's toddlers when they are adults. Our grandchildren's parents record in vivid detail almost every day of their children's lives. Each day, more than a dozen still photos and videos are usually recorded. Our grandchildren constantly view those videos and photos and talk about them. Moreover, when our grandchildren visit for extended periods, their grandmother often creates large photo books of the visit with engaging supporting narrations. For years afterward, the grandchildren pull out those books, closely inspect the images, and ask others to read the text.

With such increasingly common practices, such children may never have the feeling of infantile amnesia. By constantly reliving their experiences in photos and videos, they may rarely encounter gaps in recall. Given a universal tendency to often confuse the sources of our memories, today's toddlers may regard infantile amnesia as a mysterious defect found only in older people.

Externalization methods have proliferated in the last few decades to a degree that surpasses that of any other time in history. As a result, the consequences of outsourcing and off-loading through new technologies are constantly evolving. Viral forms of misinformation and deep fakes will create, in addition to memory differences, more distortions of our remembered pasts.[43] Traditionally, an old photograph was seen as unimpeachable evidence of a past event; now we need to be more cautious when we see alleged records of past events such as photos, including ones in which we were supposed participants.

Inadvertent Traces

Sometimes information is off-loaded in a completely incidental manner. Shortly after I arrived at Yale as a faculty member in 1998, I was asked to serve on the university library oversight committee. At that time, the Yale library's collection of books was among the world's largest and was a major budget item for the university. One of our charges was to evaluate all the library's activities in response to emerging technologies such as online search engines and online access to book and journal collections. These considerations led to an intense discussion of whether we should do away with the physical card catalogs. The full set of card catalogs consisted of row after row of slender wooden drawers in large cabinets. They occupied large amounts of floor space in grand Gothic entrance areas. Several committee members wanted to junk the physical card catalogs and switch over to remote access via terminals in the library. The card catalogs were undeniably attractive.

Their seasoned hardwood nicely complimented the building's cold stone walls, but aesthetic appeal wasn't sufficient reason to keep them. As we were about to decide, a group of colleagues, mostly from the humanities faculty, strenuously objected and asked to meet with us and explain why.

Their concerns were not primarily about aesthetics, but rather about what was conveyed by the cards themselves. Several distinguished scholars valued the unique information conveyed by the look and feel of each hand-typed card. By examining cards for frayed corners, smudges and smears, and stray markings, advocates argued they could infer how heavily that resource was accessed and withdrawn at various points in history. They even claimed it was sometimes possible to make inferences about who was looking at them. In modern parlance, prime user data was in the physical cards. As shown in figure 8.1, a heavily used card is markedly different from a neglected one. In the end we decided to preserve the card catalogs. However, they were relegated to the basement and not long afterward disappeared altogether. Everyone on the library committee was surprised that disposing of them might ruin an important source of information. My colleagues at Yale were hardly alone in their concerns. A few years earlier, the novelist Nicholson Baker had made a similar case for how much was lost when card catalogs were replaced by digital records.[44]

Searching High and Low

Advances in science and technology have a long history of liberating humans from menial, time-consuming tasks and thereby giving them more time for innovation. In prehistoric periods, advances in agricultural practices, such as irrigation, allowed humans to spend less time procuring food. More recently, machines such as the spinning jenny and combine harvesters have increased efficiencies by several orders of magnitude. People did lose old menial jobs, but these jobs were replaced by arguably better ones. Historical success stories of lives made better through technology, despite people's vociferous objections, have convinced many people to be optimistic about current advances in information technology. Nowhere is this optimism more powerful than with search engines.

Even before the current burst of AI performance, search engines, news feeds, and other improvements in accessing information were seen as enhancing our cognitive capacities. The philosophers Andy Clark and David Chalmers

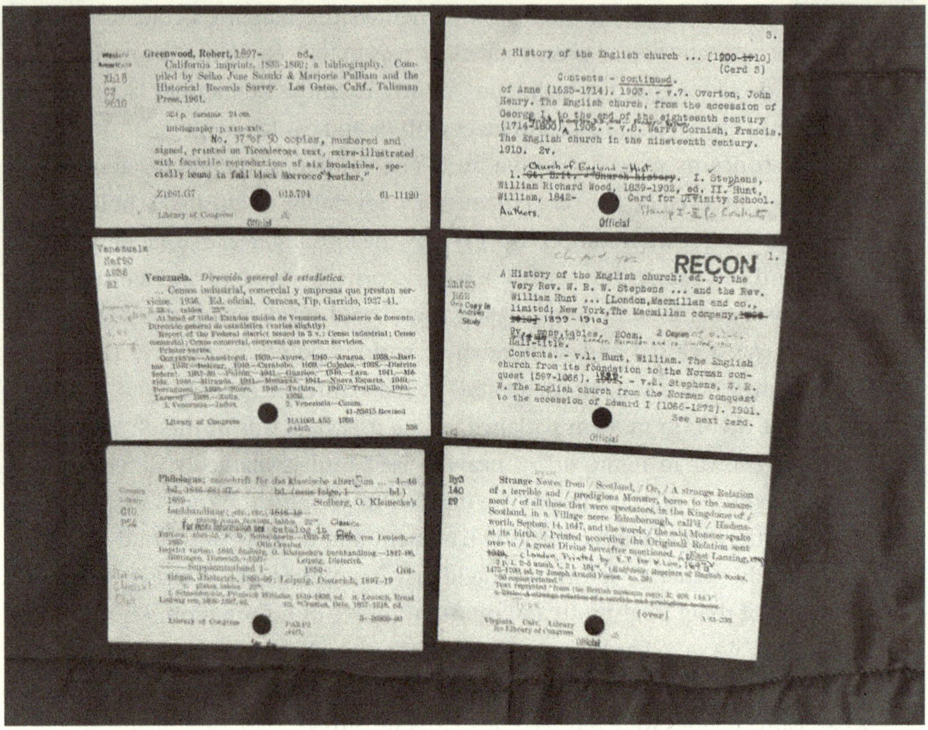

Figure 8.1
Some scholars were aghast at the thought of removing physical cards from library catalogs. They argued that valuable information about the activities of past scholars and disciplinary trends over time could be gathered from smudges, markings, and other signs of use. Compare what information might be gleaned from the cards shown here that were in Yale's master card catalog. Note, for example, that a user felt obliged to put the word *leather* in scare quotes in the upper left card. Some even argued that perfume and tobacco odors could linger on cards, providing yet more clues about users. *Source*: author.

describe how aspects of the world around us support thought by serving as extended minds. We can use our bodies to extend thought and communication through gestures; we can support thinking through specific environments such as airplane cockpits; and we can rely on other minds at scales ranging from lifelong couples to communities. All of these are aspects of off-loading and outsourcing thought.

Search engines and AI assistants might free up our minds for more effective thinking by reducing the need to access memories of information in our heads. We realize our heavy reliance on external information when we are unexpectedly cut off from all internet access for extended periods. Many people report feeling that their own mind is shrinking each day and that they are becoming less intelligent.[45] But when we have reliable internet access, this form of externally supported cognition could represent an advance in human thought. The advance might be especially large for today's children, who are growing up fully embedded in these knowledge outsourcing systems and may be much more adept navigators of different search options as a function of the kinds of information needed.

Zoomers search differently from earlier generations. Recent reports describe them as more likely to search for information on TikTok than Google—a choice that wouldn't even occur to most boomers.[46] Most people in all generations are unaware of the ways that search results can be selective and unique to individuals or to inferred groups. Search engine providers receive fees for featuring paid results intermixed with what are known as organic results, and users frequently don't notice the difference. In addition, search engine optimization (SEO) is a process through which websites try to tilt the search results in their favor by producing higher rankings than would be accurate reflections of how often a site was deliberately accessed by human users. If generations differ in how they search, people may unwittingly obtain different results as a consequence of different advertisers and optimization strategies associated with each search provider. For example, a search for "range finder basics" on Google and TikTok in 2025 yielded far more explanations of how range finders work in Google and far more explanations of how to use them in TikTok. Thus, content provided by search engines for the same query varies greatly across providers. This can easily lead to effects on explanatory insights such as emphasizing either mechanisms or functions.

Search styles and strategies change in other ways between the ages of about ten and twenty. Jonathan Kominsky and I asked how search results were evaluated by students in grades six through twelve and by young adults.[47] In addition to being interested in age-related changes, we wondered if we might also uncover hints of cohort or generational changes. In that study, all participants were told that they were to judge the results generated by a search procedure. They then saw a series of pages, each with ten simulated search results allegedly generated in response to a search query. The content domains included economics, psychology, chemistry, engineering, and biology. For each page, excerpts from the supposed first page of the website were shown, just as results were shown by search engines before mid-2025 when AI-assisted answers suddenly appeared above ranked lists of web pages in Google search results. The simulated results matched the queries in one of three different ways: they matched both the underlying disciplinary principles and the specific entity queried, or they matched the underlying disciplinary principles but not the entity, or they didn't match either the discipline or the entity. Thus, in response to the query "why do bridges need to be coated with special paints," participants received (1) a match that discussed bridges and painting and the chemistry of compounds used in specific environments, (2) a discipline-only match, which discussed various glues and focused on the chemistry of compounds used in specific environments, or (3) an entity-only match, which talked about painting bridges but focused on a volunteer group that painted bridges as part of a beautification project. All participants were asked to rate how relevant each search result was to the query.

A shift occurred during the high-school years. Younger students were more likely to rate the entity-only items as relevant and discipline-only items as irrelevant. Most middle schoolers didn't have sufficiently strong grasps of the kinds of mechanisms central to broad domains such as chemistry and biology to be able to see why information about mechanisms was more relevant to the query than mere word matches and superficial topic indicators. The shifts occurred in roughly the same manner for all five domains studied. It is not that the younger children didn't know that inside mechanisms were likely to be more useful; they just hadn't yet learned enough about mechanisms in those domains to be able to see when the principles and relations were relevant.

Shifts from focusing on surface facts and relations to considering deeper patterns and principles can occur at any age depending on typical exposure and use of each kind of information, but we can also see how differences in exposure to the two kinds of information could vary with generations.[48] Boomers, millennials, and zoomers tend to take contrasting search link journeys even when they start with the same initial query. For example, a boomer wanting to learn more about how metal detectors work might find themselves pursuing questions about electromagnetic radiation, phase shifts, and electrical conductivity. A zoomer might find themselves learning more about degrees of precision, ease of use, and the most reliable influencers for detectorists.

Perhaps today's teens have an intuitive search engine wizardry that rivals their phenomenal texting speeds. Search engine natives might be gifted at handling external stores of information. However, such enhanced skills have not yet been found. Older experts in diverse academic fields repeatedly observe how their students cannot find relevant information through internet searches. After spending years immersed in a discipline, experts learn subtle patterns concerning how phenomena cluster and how knowledge is organized around those clusters. That content knowledge is a powerful driver of effective search skills and may swamp enhanced general search abilities.

Searching Fast and Slow

The ability to quickly find information online is a form of transactive memory and has effects like those of people relying on each other for memory support. When people know that certain information is on the internet, they are more likely to forget it and focus on where to find it and not on its content.[49] Searching for causal explanations influences how we think we understand the world. Matt Fisher led a project in our lab looking for cognitive effects of searching for explanations online. We found that searching inflates people's senses of how well they understand many phenomena. For example, if asked to search online for an explanation of how glass is made or why there are phases of the moon, people subsequently think they understand those phenomena better than those who simply read the same explanation from a printed source. Even worse, they also inflate their understanding of all things that could be plausibly answered by searching online, such as why there are more Atlantic hurricanes in August and September, or why X-rays cannot penetrate lead.[50] Critically, they do not inflate their understanding of things that could not plausibly be answered online, such as how they decide what to do on weekends or why they become annoyed by some things that don't bother others.

The ease and speed with which we can find explanations online makes it difficult to tell what is in our heads as opposed to what is stored in external repositories. Finding explanations through printed sources and other older external aids makes the boundary clearer and slows us down so that we can more accurately assess our personal understanding. In the transactive memory tradition, the effect is analogous to when long-term couples overestimate what each knows in their own heads because they have become so adept at instantly and effectively filling in each other's memory gaps.[51] In later studies, Fisher showed that if the internet search is slowed down with delays or if internet searchers have to make active choices that make the boundary clearer, the inflation effect disappears.[52] He further found that when people study for quizzes by searching for material online as opposed to receiving the same material without internet search, the online searchers did worse on the subsequent quizzes while having just as much, or even more, confidence than the other group.[53]

Fast and slow searching may therefore have unintended effects on what we know and on how much we think we know. At the least, a better awareness of such effects might serve people well when they desire the most useful results. As searching becomes ever more intrinsic to each generation, we can hope that people will naturally develop a more sophisticated awareness of how different ways of searching can influence both knowledge and metaknowledge. We can also worry that as search becomes natural and automatic, its effects on what we know will become more invisible.

Just-in-Time Cognition

While search engines can be understood as external aids for the individual, the effects of search engine availability and use may be most pronounced with transactive thought. The contrast is stark to anyone who teaches college seminars. In many such seminars, students have their laptops open and their phones in view. Setting aside the many ways that laptops and phones can provide enticing distractions, even when students use them only as part of the discussion, the dynamics are radically different from meetings in which students have no access to external information and must rely on what they mentally stored prior to class. In the no-access cases, seminars have different interactive styles. The tempo of the conversation is faster, and students are more engaged with each other in real time. In one of my recent seminars, I asked students not only to make laptops and phones inaccessible, but also not to take notes. Instead, I asked them to look at each other, discuss the material, and address the key questions. Most students reported this was the first college class in which their discussions were completely untethered from external information sources and from notepad support. For students from earlier generations, learning untethered from the internet was the norm.

To be sure, untethered seminars incur costs as well. Questions arise where the discussion could be enhanced by a brief search query. Sometimes, this can be best done by asking one person to look up the information and report back to the class. Other times, having the entire class search can provide a revealing spectrum of responses. In my seminars, we consciously take breaks for brief information queries when needed and then continue as before. Even in those short interludes, if the instructor walks around the seminar table while students are searching, they will observe notifications popping up on many screens and see students attempting to covertly click on them as the FOMO overwhelms their seminar participation.

One seminar format is not always better than the other. But the norm has drifted strongly toward the fully connected classroom in which everyone is constantly checking screens and relying on external real-time sources. How we think and interact changes with these formats. We can hobble our insights if we do not effectively use those formats in ways that enrich the discussion. Both formats involve transactive cognition, but of different flavors. In the untethered versions, seminar members get to know each other's areas of expertise and areas of shared knowledge more rapidly and deeply; they learn how to rely on classmates for rapid input when needed. In the connected versions, they come to appreciate which members have the best search and selection skills for certain kinds of information and develop different, slower ways of relying on others. To do this well, they should not simply value most highly those who retrieve information the most rapidly; the quality of the information must be integrated with the speed of its retrieval.

More broadly, one inescapable contrast between the two seminar formats revolves around what we'll call just-in-time (JIT) cognition. JIT cognition is analogous to just-in-time manufacturing practices that soared in popularity in the 1980s. JIT manufacturing stressed having raw materials produced just before they were needed and in just the right amounts in order to decrease expenses associated with warehousing and stockpiling and to avoid making unreliable long-term resource demands. Similarly, JIT cognition avoids stockpiling ideas, explanations, and facts, and instead retrieves them as needed so as to avoid unnecessary burdens on real-time cognition. JIT cognition practices have soared because of the explosion of mobile smartphones enabling real-time access to almost any information, anytime and anywhere. The potential benefits are highly attractive. (A different use of the JIT concept known as JIT learning has appeared in educational research. JIT learning stresses efficiency; bite-sized, easily digestible content; and situational relevance.[54])

But consider the potential costs of JIT cognition. Just as JIT manufacturing created new problems, JIT cognition hobbles some forms of cognitive interactions. Imagine a discussion in a group seeking to increase the efficiency of electric car batteries. As members using the JIT approach come together for their first meeting, they bring their laptops and perhaps prepare by building lists of links to reliable sources on battery design and the special considerations associated with vehicles. The other group, which we'll describe as using precompiled cognition, prepares for the meeting by studying the science and engineering of batteries in situations where weight and large capacities are important. Explanatory understanding of the key issues will not emerge instantly in the minds of JIT thinkers. They must first deeply process and reflect upon the explanations they uncover through searches. The initial meetings of this group will have numerous lengthy pauses (and many early misconceptions) as members try to come up to speed on key concepts. In contrast, the precompiled cognition group will know immediately the implications of phrases such as "the anode decay problem" and its discussion will proceed quickly and efficiently. JIT thinkers are likely to heavily embrace the latest AI models that claim to be able to do the deep thinking and reflection for them; even though this takes time and requires blind trust of the AI system. This is worrisome, given the warnings from AI experts who tell us that AIs, while seeming profound, are shallow heuristic users.[55] Deep learning does not entail deep thought.

Reductions of cognitive load by relocating thought drudgery to external sources seems like a clear win.[56] But moving even mundane knowledge and skills out of our minds can incur costs when there is a clear advantage to having those elements in mind and in the moment. The hard part is knowing what to move offshore to other realms and when. Searches for information in a range of contexts are therefore likely to reveal generational differences in the content that is recovered. Initial search results in turn influence the ensuing chain of search queries. Those different search journeys could easily lead to different kinds of explanatory understandings, such as those focusing on functions versus those focusing on mechanisms.

These contrasts can be magnified as search engines and AIs start to tailor themselves for individual users. In all these cases, it is becoming ever more important to have a grasp of what one knows and where one is ignorant. We therefore now return to metacognition.

Metacognition Revisited

Metacognition, the awareness and understanding of one's own knowledge and cognitive processing, is an essential part of thinking in groups, outsourcing knowledge, and searching for information. We've seen how estimates of one's own level of understanding and learning can be inflated by internet search. Assessments of our internal knowledge can also influence when and how we search. For example, we often assess the quality and fragility of some internal knowledge in order to decide whether to tape up a note, set a reminder in a calendar, or ask a friend to remind us. We clearly do not do this for every future event. We would soon be overwhelmed by a deluge of reminders. We trust our ability to remember to turn out the porch light when going to bed or to feed the goldfish in the morning and don't leave external reminders for those things.

Children are not as vigilant about their own memory limitations and do not sufficiently appreciate when they should use external cognitive aids. They certainly rely on other people and things for learning and memory but often don't realize that they do. As we get older, we improve our linking of metacognition to outsourcing and off-loading; but we always struggle to find the optimal balances between those aspects of memory we internalize and those we externalize.[57] As new technologies appear and access to them becomes ever easier, and as younger generations become more fluent at using external aids to amplify their thinking skills, people will need to constantly refine their metacognitive skills. To rush into new technologies without an updated perspective on how one's beliefs and understandings are grounded risks the loss of more and more intellectual autonomy.

The debate on whether new digital media is making us agile, flexible wizards or lazy dullards creates a false dichotomy. New search technologies and AI-supplemented explanations can either expand or diminish our minds depending on how we understand and interact with these tools. Novices embarking on home repairs are often cautioned not to use handheld circular saws at first, and then only with great care. Major lacerations and amputated fingers are likely without appropriate knowledge of the saws and how they interact with different kinds of wood, hidden nails, and other factors unknown to most beginners. When used with skill, however, the same saw makes many carpentry tasks much quicker and easier. We may need to regard emerging information technologies as being more like circular saws and less like friendly stuffed animals. They pose major dangers for novices but can be effective tools for users with sufficient knowledge. Much of that knowledge is metaknowledge—knowledge about knowledge.

Metacognition in Groups

People use metacognitive heuristics to decide whether to accept a group's recommendations based on the kinds of the group dynamics observed and the specific task involved. Emory Richardson led a series of studies in our lab examining such heuristics. He first found that even young school children valued group opinions differently depending on the problem at hand.[58] Consider a problem requiring reasoning, such as finding the best route for a cartoon car to take through a reward-rich network of paths while never taking two rewards of the same type in a row. All ages thought the best information would emerge when a small group of people met together, discussed the problem, and provided an answer. Lower-quality results were predicted when each member of the group was asked individually and the most frequent answer was used. The preference for groups that met and discussed persisted even when those groups were compared to other groups ten times larger but who were asked for individual answers without meeting. Complex problems benefited from discussion and the exchange of ideas.

Yet participants reversed their preferences if a group was meeting to make more subjective decisions. For example, study participants were asked to decide which items from a set other people might choose as an attractive design. All ages chose summaries of the attractiveness impressions made by each individual member. They saw these summaries as obviously better than a decision emerging from group discussions.

In a second set of studies, we examined speed and accuracy trade-off tasks requiring either slow reflective thought or rapid judgments.[59] From kindergarten onward, problems involving extended thought were seen as flawed if responses were given too quickly. In contrast, perception and fact recall were seen as more likely to be of higher quality if replies were rapid. If a person is asked to say how many different combinations of people could make up three teams of eight players each where all players were drawn from the same pool of twenty-four people, a quick response would be suspicious. Figuring out the right number should take time. However, if the same person is asked whether a shirt is green or blue and takes thirty seconds to decide after staring at it intensely, we are inclined to have less confidence in his judgment. Similarly, if an individual is asked if they were present at an unusual event and they paused and seemed to think deeply before saying either yes or no, their answer is seen as likely to be of low quality. All ages showed similar patterns of judgments.

Intuitions about the contexts in which fast and slow thinking will thrive require sensing how different aspects of both perception and cognition relate to processing speed. Integrating the relevant information is an impressive skill at any age. This second set of studies links to the first set when we consider fast and slow thinking in a group context. In cases where judgments come to mind immediately, input from others in your group might degrade judgment quality as factors such as social pressure and conformity effects begin to hold sway. With slow reflective judgments, the influence of the group may not be as negative.

From the first years of school onward, people are sensitive to how mental states in groups influence problem-solving and decision-making. Thus, in addition to mastering the complexities of finding common ground, we learn early on that thinking in groups can be highly effective but only when specific cognitive conditions are met. As members of several overlapping communities, we bring all these intuitions to bear so that we may derive the most benefit from thinking in groups.

Zoomers' interactions with others could also have a different tempo. Zoomers may be constantly looking for relevant information from online sources and thereby access information more quickly; but they may often lag in face-to-face meetings as they disconnect from the in-person discussion to find information online. Even if they do not disconnect fully, they may be tracking the conversation at a broader but shallower level (such as noting positive or negative tone) without processing arguments, models, reasons, and evidence. Both zoomers and boomers may use response time as signs of response quality, but they might differ on when fast and slow responses are seen positively or negatively. A boomer in a cargo ship design team might see a rapid response on the best hull shape as indicating inadequate consideration of a complex set of trade-offs between such things as double versus single hull designs. In contrast, a zoomer might be impressed when a group member is the fastest at accessing an article stating which design is best. For this content area, zoomers may put more value on skills relating to quick online retrieval than on skills relating to slower reflection on relevant causal models that take considerable time and prior study to understand.

Coming Together

Advances in technology can change how we think and what we think about. With GPS guidance nearly everywhere, we follow turn-by-turn instructions much more and use internal cognitive maps much less. Zoomers create and share videos instantly in ways that were inconceivable just a few decades ago. Elaborate new etiquettes have appeared regarding texting, phoning, emailing, and other ways of communicating. Zoomers happily list activities they now think are outdated and of little use: reading maps, writing cursive, and using copy machines. New communication technologies have always elbowed aside their predecessors, whether it was the printed word and literacy, stained glass window stories in churches, telegraphs and timely news, or radio broadcasts of sports. But the breadth, depth, and rate of change all seem greater today. The contents of our mental lives are clearly changing. The psychological consequences of those changes are less clear.

With respect to social and emotional development, the new digital media is widely seen as having devastating impacts on many teens and young adults, especially when they come together in groups. In contrast, there is little agreement about its effects on cognition. Is cognition more resilient across different environments? Such a conclusion would not fit with decades of work claiming that everything from pandemics to preschool can have powerful influences on cognitive performance. Instead, the differences may lie in what cognitive effects are studied. The primary focus has been on general cognitive skills involving attention and memory. But few studies have examined the effects of changes in broader content areas. What happens when different generations come to know different kinds of recurring relational patterns? Zoomers have encountered far less information about how and why things work. The perceived importance of underlying mechanisms has largely evaporated. It may have been replaced by procedural and functional knowledge and new abilities to rapidly access information services.

The most sensible default position is to assume the effects of new technologies on thinking in groups will be both positive and negative and to then consider the costs and benefits in each situation. Most views tend to be strongly negative or positive and to ridicule the opposite accounts. For that reason, we can appreciate commentators who take more nuanced perspectives. For example, the writer Clive Thompson, while being largely positive, thoughtfully acknowledges losses in areas involving more reflective thought. He tends to think most losses are being displaced by more powerful and effective ways of thinking, often just because they are transactive; but he admits that some valued forms of reasoning and idea generation may decline.[60] He considers at length the case of writing, especially in emerging transactional settings. While many have bemoaned a perceived decline in the writing of long classically structured papers, Thompson sees this as shortsighted.

When we survey all forms of written content on the internet, zoomers are writing up a storm. They are outwriting not just their parents and grandparents but even those legendary letter writers of the Victorian period. Thompson sees zoomers as becoming better writers than older generations because they're writing in so many different forms and venues. Their writing is abundant in texts, blogs, captions, and many other formats. He sees young writers as more agile and flexible in their use of the written word. They are more likely to engage in sophisticated debates and arguments and flexibly take different points of view. Yet he also values long-form writing that emerges out of intensive individual nontransactive work, such as laboring over a tightly argued review. All of us who care about this general topic—and that should be everyone—must keep in mind the gains and losses accompanying any new form of technology. Just as the latest thing is rarely a positive development in all respects, we need to consider carefully what we might be losing in terms of skills and perspectives. When we are inclined to be wholly negative or positive, we should also work hard to see if a more balanced position is closer to reality. We certainly don't have to always end up with bland equivocating conclusions, but it seems rash to not evaluate both costs and benefits outside our personal experiences.

* * *

We have considered how the changing world of information availability might have distinctive consequences for how older children and adults think in groups. Synchronous and asynchronous interactions draw on knowledge and skills in different ways and therefore show different effects on cognitive content and activities. The sudden proliferation of mobile phones and then smartphones has resulted in people spending far less time generating cognitive maps and solving a large class of coordination problems that were routinely addressed by prior generations. Given extensive findings of brain changes in the hippocampus in those who use cognitive maps intensely, such as older London taxi drivers, we should not underestimate just how much these different experiences might shift how cognitive and neural resources are allocated in the minds of adolescents and young adults.

As modes of communication have proliferated, younger generations have spontaneously developed ever more elaborate rules of digital etiquette. These rules may further develop into a different set of cognitive skills as knowledge of underlying causal patterns declines. This is arguably adaptive and appropriate. If complexities in the sciences are starting to exceed the cognitive grasp of more and more people, perhaps they need to direct their thoughts elsewhere and learn more about how to rely on others.

Building on the findings of research on transactive memory, a broader class of phenomena is emerging in interactions between individuals, which we have described here as transactive thought. The rise of artifact complexity many thousands of years ago in prehistory illustrates the power added to individual thought when it is embedded in a culture. Cultures have always amplified thought in individuals and built on the achievements of prior generations. However, the pace of cognitive enhancements outside individual minds has surged in the past few decades. If mechanical knowledge is an especially well-tailored fit between an individual's mind and the world, the declining quality of that knowledge may not be easily compensated by other skills.

One major benefit of transactive thought, broadly construed, is cognitive outsourcing and off-loading, which seeks to optimize the storage and processing limits of human minds by reallocating loads to external minds and devices so as to free up internal capacities. But we may regain those capacities at considerable costs, especially in real-time group interactions.

Transactive cognition heavily involves metacognition and deference. These two skills are intertwined with how we think about human explanations and what their limits are. It is time to return to what explanations mean for humans. Chapter 9 asks whether the recent evolution of everyday devices into unexplainable entities represents the first wave of a more massive AI-fueled shift to the unexplainable.

9 The Unexplained and the Unexplainable

Science is often depicted as a progression of discoveries of mechanisms, principles, and models. These discoveries explain what lies beneath and what is responsible for long-observed regularities. Making such discoveries can take a long time. People have puzzled over some regularities for millennia before long-sought explanations finally came to light. Prehistoric humans noticed how the phases of the moon covaried with the tides, but not until Newton's model of gravity did an explanation emerge. Ancient humans knew that water expanded when it froze but olive oil did not. But only in the 1930s did Linus Pauling show how the special properties of hydrogen bonds in water yielded a causal mechanism. These cases of finding elegant explanations after centuries of attempts have been interpreted as signs that explanations for everything will ultimately be revealed. When the rate of discoveries accelerated during the Enlightenment, people started to believe that all corners of the natural world would soon yield to explanations for human comprehension.

But some patterns have resisted explanation. Even the most pervasive and long-standing correlations can remain mechanistic mysteries. One such pattern of correlations is the *latitudinal diversity gradient* (LDG).[1] The LDG is straightforward: The closer to the equator, the more diverse the range of species in that latitude. The LDG holds all over the world, and according to fossil records, it has held for as far back as biologists and geologists can measure. Yet, this robust correlation remains unexplained despite vigorous attempts by many leading researchers. Several different theories have emerged, but all have serious problems, and none have garnered a consensus.[2] Inspired by Alexander von Humboldt's revolutionary view of nature that arose from noting adaptive similarities between plants and animals at different altitudes, scientists found an altitudinal diversity gradient

(ADG) very much like the LDG. This led to searches for a unified overarching explanation of both the ADG and LDG; but that explanation has not appeared. The posited mechanisms for the ADG and LDG are markedly different. The LDG is assumed to arise from long-term ecological and evolutionary factors working across immense geographic and climatic ranges. The ADG is assumed to occur because of local, relatively rapid changes in the environment over relatively short time scales. These contrasts have thwarted attempts for unified accounts of the LDG and ADG phenomena in terms of a common mechanism. The two gradients are depicted in figure 9.1.

These differences in scale, environmental changes, and ecological dynamics mean that while some principles of biodiversity gradients may apply to both the LDG and the ADG, they will always require distinct explanations

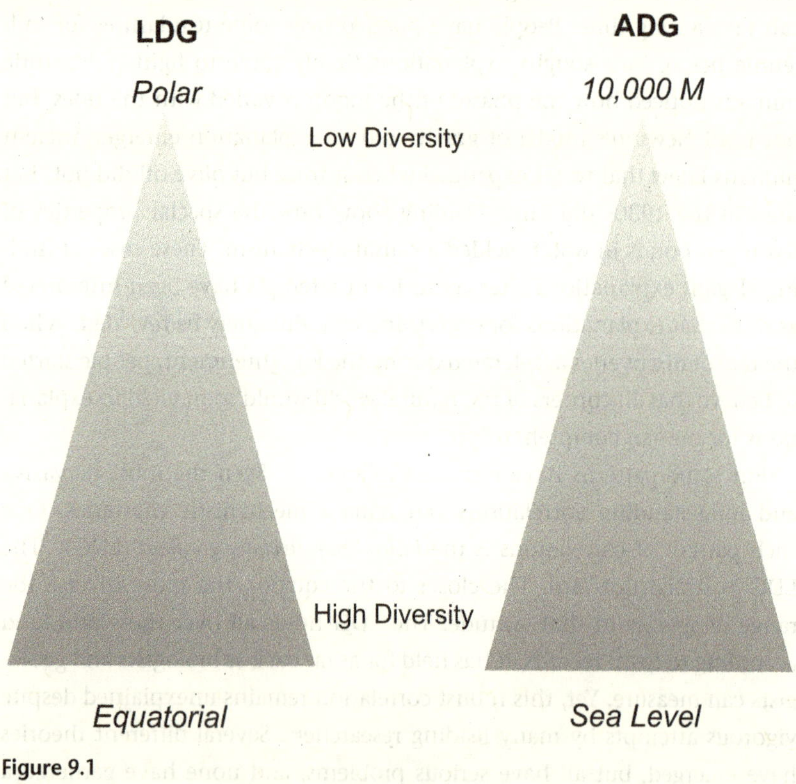

Figure 9.1
The LDG and the ADG seem strikingly similar. Yet there is no consensus on the underlying mechanisms for each. There is complete agreement, however, that the two types of gradients must work through different mechanisms. *Source*: author.

and mechanisms to fully understand the patterns observed. Many scholars now view the LDG as a case where an interactive cluster of several mechanisms is at work—and this Gordian knot of causal threads might never yield a comprehensible explanation.[3] Others remain hopeful that an elegant mechanistic model will finally emerge. Regardless of how the LDG debate is resolved, it illustrates how some strong correlations in nature, while resulting from causal processes, may never be easily explainable to human minds.

Recurring claims of the end of science partially reflect a growing tendency to abandon hope of always finding straightforward, comprehensible mechanisms.[4] Claims that science has already picked all the low-hanging fruit may be equating low-hanging fruit with finding cognitively comfortable explanations. But science also involves making precise predictions, and in many areas, the ability to make more accurate predictions is soaring even if insight into mechanisms is disappearing. This shift in emphasis is driven by a revolution in computational methods involving deep learning and related processes. Improvements in performance have reinvigorated the field of artificial intelligence. When AIs provide improved predictions at the cost of not knowing how they are making those predictions, those areas of unknown processing are called black boxes. The expanding black boxes of mystery mechanisms seem eerily like the decline of mechanistic transparency in devices discussed in previous chapters. Many causal patterns and correlation networks may not be captured by comprehensible models and mechanisms. The intractability of turbulence to useful models and mechanisms has long been treated as one of the few outlier cases in an otherwise orderly mechanical world; but that view may have been more of a reflection of our tendency to look where the lights of insight were bright and to neglect what lies in the shadows.[5]

We have seen how the world of explainable things can shrink as devices become undecipherable blocks of silicon that are impossible to reverse engineer by examining them with the naked eye. But those devices are not intrinsically unexplainable to everyone. Experts know the precise circuits that make toasters, timers, and GPS units work; they can describe the mechanisms in full detail (or at least they can do so as a transactive group). Even when it is impossible for laypeople to divine those explanations through inspection and tinkering, they can learn those explanations if they connect with appropriate specialists.

To obtain stronger predictions through new computational techniques, researchers across the sciences are sacrificing insight into mechanisms. AI performance has improved because of larger datasets and improved algorithms for inferring networks of probabilistic relations. Going under the hood of a deep learning program reveals an alien terrain with enormous numbers of parameters and their interrelations. Deep learning doesn't just look at interrelations between inputs such as a dozen properties in each pixel of a photograph; it infers new parameters from those inputs, which are then implemented at multiple levels of processing. The parameters and their links are unlike anything we experience in our own minds. The problem of recovering an explanation is immediately obvious. Humans might examine a dozen or so factors present in each pixel of a CT scan and then ask which factors or simple combinations of them best predict a form of cancer. A deep learning program might achieve greater accuracy than experts; but those gains in accuracy require creating many thousands, or even billions, of layered parameters and then tracking fantastically complex interactions between them. Many of those parameters are uninterpretable to humans, and their interactions are even more enigmatic.

Not knowing how deep learning arrives at a prediction is troubling not only because the explanatory gap is conceptually uncomfortable, but also because of the possibility of unanticipated errors. Early warning systems of nuclear attacks must make split-second decisions about large numbers of inputs such as satellite reports of remote flashes of light, vehicle movements over two weeks, and cloud cover. If AIs can detect missile launches more accurately and rapidly than humans, we might want to use them. Yet, as detection accuracy and speed improve, the underlying steps become murkier, and serious flaws may be missed. Glitches in early versions of such systems are well known. A 2018 facial recognition program by Amazon matched photos of twenty-eight members of the US Congress to people with criminal backgrounds. As best we know so far, none were actual criminals. In 2023 a Microsoft chatbot tried to convince a New York Times columnist that he did not love his wife and instead really loved the bot.[6] Detecting such flaws is ever more challenging as system complexities rise. Misleading glitches still exist in current AIs but may remain opaque until they lead to an obvious error. More subtle errors and biases may never be found.

While continued improvements in AI make some of the earlier mistakes less likely, other failures continue. Self-driving cars continue to run red lights; social media sites mistakenly flag and remove legitimate content; fake reviews proliferate on shopping websites; algorithms for cropping and cleaning up faces have racial biases; health monitoring software provides incorrect readings; hiring tools show bias against demographic groups; AI-assisted processing of medical forms leads to mistakes and delays for patients. We still cannot fully trust what artificial intelligence systems tell us.

Whether the task is related to national defense, medical diagnosis, or crime detection, AI performance is becoming ever harder to understand. As a result, the area known as explainable AI (XAI) is one of the fastest growing areas of AI research, with several thousand articles on the topic appearing in the last few years. Moreover, in light of the rising challenges in grasping how such systems work, some commentators feel licensed to wonder if the latest AIs have souls or moral rights.[7] Evaluating such ambitious claims about AI is difficult without knowing what is happening within the layered networks of parameters. These issues raise questions about generational changes in explanatory preferences. Will zoomers and later generations view the XAI problem in different ways? Or will they not even agree that there is an XAI problem?

Research on explanations in humans may help us meet these AI-driven challenges of explainability. More refined views of ignorance could lead to strategies for handling these new ways of not knowing. For example, the natural nesting of levels of expertise in societies could provide clues for navigating new forms of artificial expertise and for deciding when to outsource expertise to an AI. The ways that younger generations are adapting to disappearing mechanisms in devices may hold clues for how to adapt to missing mechanisms in the sciences. At the same time, we must consider asking more of AI systems to enable human users to trust them more fully. To address these issues, we first examine how science is being affected by the recent surge in AI.

Science Without Insight?

Many areas of science are moving toward explanatory opacity. Scientists are divided as to whether this movement is a refreshingly pragmatic perspective or a dangerous surrender of understanding to meaningless, but ever more precise, predictions. Wherever there are large datasets, understanding may be sacrificed at the altar of prediction. As those datasets grow exponentially, thanks to an explosion of new sensors for everything and a proliferation of interpretations of their outputs, big data is becoming an irresistible way to refine predictive accuracy.

Consider the seemingly straightforward question of predicting a person's height given the height of their parents. Tall parents tend to have tall children and short parents have short children. Environmental differences can influence height as well, but in comparable environments, genes rule. Genetic variation accounts for roughly 80 percent of people's height variation. We can therefore predict height quite well in comparison to many other traits, such as personality differences. It might therefore seem that explaining the mechanisms behind differences in height should be easy. Perhaps a single height gene regulates a single skeletal growth hormone and a few variations in nucleotides called SNPS (single-nucleotide polymorphisms, pronounced "snips") account for all differences in height. Yet the still unfolding account is nothing like a simple pathway from DNA to height.

At last count, at least 12,211 different SNPs contribute to accurate predictions of height. Leave out a SNP and predictive accuracy drops. A 2022 paper in *Nature* described this number as reaching "saturation."[8] With saturation, additional SNPs do not improve predictions of height variation. Given that saturation claims have been made before, with smaller numbers of SNPs, the final number may be closer to around 15,000, especially when a broader range of populations is studied; but the point is clear. Mendel's one-to-one mappings between gene variations and trait variations are incredibly rare in comparison to the complex interactions between genes and parts of genes. The large number of authors of the *Nature* paper is also telling. It is difficult to count the number because groups are listed as well as people, but more than six hundred names are listed in a long scroll of the author list. The enterprise is far different from that of a lone monk puttering in a pea garden.

Similar trends are emerging in many other research areas unrelated to height. Cognitive neuroscience is one notable example. Researchers seek to identify the brain circuits responsible for different psychological states and processes. How do processes in the brain give rise to having a certain concept or feeling a specific emotion? Brain activity levels in a set of small brain volumes known as *voxels* are correlated with each other over specified time intervals. Each voxel may have 750,000 to 2.5 million interconnected neurons and 8 billion to 20 billion synapses. Those correlations are then used to construct a three-dimensional map of brain circuits. This approach is known as *multivoxel pattern analysis* or MVPA. The size of each voxel varies as a function of the resolution of fMRI scanners such that a whole brain might be decomposed into 100,000 to 1,000,000 voxels. Large sets of those voxels are then tracked simultaneously in terms of activation patterns. These scans generate enormous datasets requiring large computer systems and sophisticated processing.

As with tracking genes for height, the more voxels included, the better the prediction of a specific cognitive process. Researchers try to reach correlational "saturation" with the smallest number of voxels possible. They hope that the saturated set of voxels will reveal a well-defined circuit, but unambiguous and revealing circuits remain elusive. In addition, voxels outside the circuit of interest have an annoying tendency to make predictions a little bit better when they are added to the mix. Again, this extra-circuit activity comes at the cost of explanation as mechanisms start to disappear in a miasma of MVPAs.

To make sense of the torrents of data comprising inputs like SNPs and MVPAs, massive computational power and technology is required. But that technology starts to take on a life of its own, as it does in AI research. Across all the sciences, the quest for more accurate predictions is turning explanatory insights into murky mysteries.[9] The XAI problem and related problems in the sciences have become a complex and challenging literature for those without extensive knowledge of state-of-the-art deep learning. The leading experts often disagree or confess to not having a solution. Reviews attempting to construct typologies of different XAI approaches come up with different categories and clusters; many describe the accelerating pace of new learning programs as making typologies obsolete within weeks of their posting.

The XAI phenomenon is clearly different from vanishing physical mechanisms. But in some respects, young children's experiences relate to commonly reported XAI dilemmas. As a child grows older, feelings of certain things being unexplainable may morph into the sense that those things are explainable even if the child doesn't know the explanation.[10] Repeated experiences of this sort in development should help adults distinguish and cope with different kinds of explainable and unexplainable things. Useful strategies may result from comparisons of how human minds learn causal patterns with how AI systems make predictions about the same patterns.[11]

Promoting Prediction

As deep learning models make ever more accurate predictions in the sciences, a growing group of researchers has come to explicitly prefer prediction over theory. One prominent example refers to the incidence of bold but incorrect theories and explanations for mental disorders. These initially seductive, but ultimately fatally flawed, theories have led to theory aversion. A widely cited essay was entitled "Choosing Prediction over Explanation in Psychology: Lessons from Machine Learning."[12] Not surprisingly, others have pushed back and argued at least for blended models that incorporate both explanation and prediction.[13] That balanced approach to psychology seems less radical and disruptive; but its future will depend on the ability to build reliable hybrid systems—a challenge that remains unmet.

A focus on outputs and not on causal models or mechanisms is hardly new. The history of early antibiotic research was largely driven by the goal of finding more effective antibiotics by means of massive atheoretical searches for new fungi in soils. Soil samples were drawn from every possible source. These ranged from ocean floors to arctic ice, mortuaries, garbage dumps, and mine shafts. At one point in the late 1940s, Pfizer had roughly 135,000 distinct soils samples.[14] Similar frantic hunts for better variants with no motivating theories or mechanisms continue everywhere from pharmaceutical labs to superconducting materials research centers. In many cases, such approaches have been the best ways to proceed, especially when there are easy ways to obtain and test different inputs.

Current approaches to science may offer a more principled objection; in many instances, complete causal mechanistic models may never exist that fit with our cognitive limitations. In prior decades, those who focused on outputs and results were not against theory and mechanisms. They just thought that mechanisms would come later. Most researchers assumed a future with mechanisms and models would eventually be the fruit of their initial trial and error. They believed that knowledge of mechanisms would be essential to address novel problems and to make dramatic innovations. Indeed, the COVID-19 pandemic might have been worse without mechanistic models of mRNA driving vaccine design and synthesis.[15] If such models are becoming rarer, novel threats may be more difficult to address through even the largest statistical datasets and deep learning systems.

In technology sectors, many traditional engineering approaches started with goals of incremental improvements to a specific device; but they also embraced models and mechanisms. Engineers saw the two as going hand in hand. Mechanisms constrained an otherwise overwhelming set of choices on what to tweak in current designs.[16] Even with entities that cannot be fully described by discrete models, engineers use more localized mechanistic models to simplify their design tasks. For example, designing new aircraft involves tinkering and informed guesswork about wing shapes combined with precise mechanistic models of components.

As datasets proliferate and as computational power increases, improving predictions becomes an ever more seductive goal. Improving predictive accuracy seems more straightforward than discovering new mechanisms. In medicine, the migration toward purely predictive approaches is accelerating as researchers use deep learning to diagnose diseases and predict the outcomes of disease treatments. Deep learning is also behind much of the new emphasis on "precision medicine," where a patient's specific genome drives an individualized treatment plan.[17] Increasingly, doctors are finding themselves following machine learning recommendations tailored for each patient without knowing the decision processes supporting those specific recommendations. As a result, they are becoming less able to explain why they are using a particular treatment plan.

Several climate-change research programs have retreated from detailed mechanisms and instead focus on obtaining more accurate predictions from expanding sensor data.[18] Econometric models have frequently become so massive in terms of their datasets that prediction seems to be the only feasible way forward. Predictions of crop yields and disease outbreaks use datasets that include weather patterns, soil conditions, and pest populations. Once again, oversimplified but humanly conceivable models may do more harm than good. Urban planning is yet another example; researchers are attempting to factor in data from social media, traffic cameras, energy bills, migration patterns, and birth rates, among other sources.

Given the history of failed theories and weak explanations, the rise of anti-explanation proposals is understandable. But potential costs must also be assessed. When AI models suggest actions or give responses generated by their impenetrable black boxes, we need ways to evaluate risks and benefits. Attempts to describe weak areas of AI systems are especially difficult as they are rapidly evolving moving targets. What, then, are feasible approaches to managing risks?

In many instances, AI recommendations on specific issues can be compared to those of human experts. When the two sources differ significantly, more exhaustive analyses of the problem are warranted, even if the AI alternative has a stronger track record. Many AI systems may show the largest departures from humans in tasks requiring generalizing to new circumstances. Rare cases that don't easily fit into neat problem categories also are likely to reveal contrasts between humans and AI recommendations.[19]

Hopeful Hybrids

Worries about prediction procedures completely displacing models have led to proposals for hybrid systems containing both. While this strategy is appealing, there is no consensus on how such hybrids should be structured. Some blends have begun to be used in molecular biology. One review suggests that predicting 3D protein structures from chains of amino acids has shifted from being computationally intractable to providing impressive results driven by deep learning. The authors suggest that, while unexplainable areas remain, deep learning has suggested important new property clusters and causal relations to humans.[20] The emergence of new causal relations may validate the continued utility of humanly interpretable models. The same review admits that these promising developments still leave huge mysteries, however, such as not yet knowing what proportion of the AI's successful predictions will remain persistently unexplainable.

Even with those cautions, many researchers across diverse fields are optimistic about the promise of hybrid architectures as well as approaches that try to integrate several different AI systems.[21] However, based on studies in our lab and several others, humans are well known for thinking they have stronger explanatory understandings than they really do.[22] People may feel that they have gained explanatory insight when they haven't. Similarly, users of AIs may develop illusions of explanatory depth, as well as comparable illusions of explanatory breadth and of objectivity.[23] We can reduce these illusions in humans by asking them to elaborate on their assumed understandings and thereby reveal gaping holes they had overlooked; but it is not easy for people to constantly check the quality of their own understandings.[24] It is even less obvious how to build AIs that check for their own illusions of explanatory depth, assuming such self-deceptions exist in LLMs and other complex systems. A different wrinkle concerns known and unknown biases in artificial and human systems. In AI, some biases, such as those guiding induction, are built in by design to enhance performance; these apparently differ from natural human inductive biases.[25] Thorny issues remain concerning how to integrate the two biases in hybrids.

The nature of AI's involvement in science remains under active debate, and proposals vary widely across the sciences. At least for the near future, blends of human cognition and AI will continue. That prospect highlights a need to adopt "meta" perspectives asking what is distinctive about AI and human approaches. One rule of thumb we've discussed is to pay extra attention whenever human and AI results diverge. A similar rule would be to increase vigilance when different AIs diverge. With disagreements, either AIs or humans may be more correct, or both may be wrong to the same degree; but we should embrace the practice of tackling important problems from different perspectives and carefully examining agreements and disagreements. When problems are rigorously examined with different approaches, the distortions introduced by each approach tend to be factored out and enable us to see more clearly what is closer to the truth. The idea of different perspectives needs elaboration as well. One approach could compare AIs in terms of the techniques they employ, such as ones that heavily use gradient-boosted trees versus those that do not. Another approach could compare data sources, such as predictions favoring peer reviewed sources versus the most recent sources, or sources from North America versus Asia. The set of possible comparisons is enormous and would benefit from systematic accounts of which ones are likely to be most informative and why.

Scientists have always worked with black-boxed mechanisms. Many cognitive neuroscientists use fMRI to explore neuronal activity in human brains but may have little or no grasp of the details of how an fMRI machine works. Biologists may use cryo-electron microscopes to explore cellular properties in tissues and not know how these devices manage to be so superior to traditional electron microscopes. In these and many other research projects, large parts of the inner mechanisms of their instruments are unexplainable to the researchers. But this type of being unexplainable differs in one critical way. These devices are not unexplainable to everyone. Moreover, bench scientists usually have a good idea of how to access relevant experts when needed. Here too, however, they may sometimes assume that experts have full understandings of why something works when in fact they don't and are still puzzling over why the device works so well. Scientists who use tools they don't fully understand also often assume that the details don't matter. If a newer electron microscope provides higher resolution, a lab might eagerly start using it; but without knowing its new mechanisms, the researchers might not appreciate how distortions in the synthesized images could lead to mistaken inferences about atomic structures.

Concerns about a future science of prediction without explanation go far beyond practical benefits. Even if predictions by AIs were universally better and always led to better outcomes, we may feel an increasing discomfort of not knowing what lies beneath. We have all experienced the joy of looking at phenomena with mechanistically informed vision. We don't want that pleasure to be based on incorrect mechanisms, but if science focuses on predictions without accompanying experiences of insight, we all may envy our forebears who reveled in the discovery of whirring microstructures behind what was around them.

Causal Parroting or Causal Primacy?

As prediction-focused research has blossomed, it is often accompanied by increasing skepticism about the use, and even the reality, of causation and mechanism constructs in many AI systems, and possibly even in humans. What does it mean to think causally, especially in the physical mechanical senses that have been so prominent in the history of science and engineering? While traditionally, causal mechanical models have been touchstones for building explanations and making predictions in many areas of the sciences, those clockwork models are losing adherents to groups who favor newer senses of causality or who abandon the notion altogether. This can be an appropriate and adaptive shift that resonates well with new AI-supported perspectives of cause. But abandoning mechanistic models may also remove powerful conceptual tools for human understanding and innovation.

The debates over the relations between patterns of correlations and causation are complex and continuously evolving, with hundreds of recent technical publications appearing each year in philosophy, computer science, psychology, statistics, and the natural sciences.[26] Here we focus on links to vanishing explanations.

Association and Causation

Roughly twenty-five years ago, the computer scientist Judea Pearl proposed the idea of a "ladder of causation."[27] His proposal helped organize ensuing discussions of what it means to process information in a causal manner. The ladder had three rungs, which should sound familiar at this point: association, intervention, and counterfactuals. Pearl suggested structured relations between these rungs. The association level, which is sometimes described as *seeing*, is the bottom level. It identifies and tabulates associations between variables. It consists of observing data and tracking correlations. The intervention level, which is sometimes described as *doing*, focuses on interventions associated with different outcomes and shifts the emphasis to causation, going beyond mere correlation. At this level we can predict the effects of specific actions or making changes in some property of a system. This kind of prediction is at the heart of true experimental designs in the sciences. The top level, counterfactuals, which is sometimes described as *imagining*, involves inferring outcomes from hypothetical situations that have never happened and may never happen. It involves understanding how causes and effects relate in other possible worlds. It asks what-if questions and involves thinking about alternate realities.

Pearl's ladder and the idea of higher rungs building on lower ones has implications far beyond three simple divisions. For example, the interventions level requires more than simply noticing and recording a correlation; it requires an ordered relation in which acting in a particular way (e.g., throwing a switch) results in a specific change (a light comes on). Here is where cause enters and goes beyond association and correlation by implying that causes precede effects in time. The counterfactual level involves interconnected sets of causal relations implying models or mechanisms. With those in mind, we can construct alternative realities and even worlds that couldn't exist. We also make apparently sensible causal inferences and predictions about imagined events in those worlds.

This quick summary glosses over vital details such as how to decide among a seemingly infinite set of potential correlations to associate. If you are dropped into a strange situation and want to make sense of it, what do you track among such properties as colors, shapes, sizes, weights, textures, velocity, movements, odors, accelerations, temporal delays, temperatures, fragility, and elasticity? Each of these could be unpacked into hundreds of more nuanced sub-properties. In addition, one must decide on what size time windows to use to bound correlation tracking. Similarly, as noted earlier, how do we decide what feature or features to change in counterfactual reasoning? Answers to these questions are under intense discussion across the cognitive sciences.

Several other approaches try to systematically contrast correlation and causation.[28] This line of work hopes to uncover potentially critical contrasts between human and artificial intelligences. Humans, even infants, immediately interpret the world in terms of causes and effects, and arguably in terms of clockwork mechanisms. But AI systems have not traditionally been described as having causality as an intrinsic part of their architecture. This contrast has led to ongoing debates such as whether AIs never really "get" causality but are merely "causal parrots" that may use causal words but do not understand what they mean beyond correlation patterns.[29] This limitation is seen as hobbling AI systems, especially when they are confronted with novel situations where prior large datasets are not helpful but reasoning from causal models and mechanisms is effective. Similarly, when detailed tabulations of all possible covariations overwhelm even the most powerful computers, qualitative reasoning with causal models may enable faster problem-solving. Causal models may also enable generalizations across different cases and contexts in ways that correlations cannot.

Critics of the enriched view of causation worry about systems that go too far beyond the information given. They cite examples of how runaway theories and models can trample valuable correlational data and result in caricatures that ignore critical information. They may further argue that when correlations are sufficiently nested in structured sets of relations, such as hierarchies, particular patterns of nesting are uniquely associated with causation but do not require it to be a conceptual primitive. In this view, causality emerges as a byproduct of how structured relations are interpreted, rather than as a primitive notion directly represented in thought. Finally, a third view proposes that AIs be supplemented with explicit models and mechanisms that are built in from the start, enabling them to always have basic causal models.[30]

Whether such hierarchical patterns are more like true causal reasoning or parroting is relevant to how AIs learn causal patterns and understandings and how their learning progression does and does not resemble that of humans. With humans, progressions can be from infancy to adulthood or from novice to expert states. We don't need to make computers think just like us; we wouldn't want to if they can do some things better. But we do want to know how they think and learn differently from us and the implications of those differences for understanding their recommendations. We want to understand these learning trajectory differences and their consequences even when we can't fully grasp what is going on inside their black boxes.

Symbolic and Sub-Symbolic Thought

Closely related to correlation/causation controversies are broader discussions in cognitive science of symbolic versus sub-symbolic thought. Symbolic thought manipulates symbols and relations between symbols in a language-like system in the brain and is often called the *language of thought*, or LOT.[31] The LOT is not simply having a spoken language in the head, and recent evidence suggests the presence of a LOT in the minds of human infants long before they have learned language.[32] The LOT is usually assumed to include a deductive logic system containing concepts like *and* and *or* and the rules of deductive logic, such as modus ponens. Proposals for cause as a conceptual primitive tend to see it as a core element present in infants' LOT and not reducible to statistical patterns.

Sub-symbolic thought occurs without explicit rules or conceptual primitives. Instead, it relies strictly on information gathered through statistical learning or associations between primitive nonconceptual features such as simple visual features. It may infer something like cause and causal relations from nested sets of correlations, but it may never explicitly represent the concept *cause*. Cause is merely a certain activation pattern representing a network of correlations. In humans, statistical learning, and presumably sub-symbolic thought, usually happens outside explicit awareness and can occur at several levels (e.g., lines, angles, geometric figures) at the same time. When people try to consciously attend to their own statistical learning, their performance often suffers.

While some may try to relegate the LOT and the concept of cause to the realm of convenient fictions, most scholars acknowledge the presence of both symbolic and sub-symbolic thought at least in the minds of older children and adults. But if both are present, how do they interact and form an integrated system? Those who see cause and other conceptual operators as present early in development struggle to find points of agreement with others who see cause and conceptual operators as emerging late. With respect to AI systems, researchers are exploring the trade-offs associated with building in such conceptual primitives, such as core knowledge of physics and psychology, in deep learning systems' architectures as opposed to having them gradually learn better approximations of core knowledge.[33] Knowing how such choices have been made for a particular AI may give humans some sense of why that AI makes its choices.

Explanatory Ignorance

One approach to making AI more explainable involves refining our intuitions about ignorance. Knowing what kinds of ignorance are present in attempts to understand an entity suggests what is explainable and unexplainable and why. We may also learn more about what strategies best handle different types of ignorance. Across the disciplines, ignorance has been described in many ways. The historian David Burke lists fifty-nine "varieties" of ignorance while assuring the reader that the list is not complete.[34] The notion of explanatory ignorance helps narrow the set down to the types of ignorance relevant to this discussion. We'll define explanatory ignorance as not knowing explanations as opposed to not knowing facts.

For explanatory ignorance, consider an easy-to-learn but commonly unknown explanation: how a manual pop riveter works. Squeezing the lever pulls back a pin (the mandrel), which forces the rivet body to bulge out on the opposite side of the hole from the riveter. Seeking explanations of how things work mechanically is constrained by our notions of mechanical relevance. When asking how the riveter causes the rivet to change, the question is how the inside parts of the riveter functionally change the rivet. One could conceivably instead focus on an irrelevant unknown causal relation between riveters and rivets, such as asking how smashing the rivet with the riveter changes the rivet, but nothing about the two parts and their affordances suggests that such an action is a reasonable way to fill the explanatory gap. There is more to discover, such as how the riveter grabs the mandrel in the rivet, pulls it up to flatten the rivet end, and then neatly clips off the mandrel. The metal of the rivet itself is also important. Aluminum rivets are easier to flatten and require less pulling on the handles than stainless steel rivets. Filling in a bit of explanatory ignorance often opens new areas of explanatory ignorance in ways that filling in factual gaps does not.

When we are curious about an explanation that no one currently knows, we might still think it will be discovered. (Except for rare cases of stumbling across artifacts that no one has seen before; these cases almost always involve natural science phenomena.) We may ask how sea turtles manage to return to a pinpoint location in the open ocean and find out that no one yet knows the full story; we just have some suggestive fragments. But most people who know about this problem feel the full explanation will be uncovered and will be understandable. Moreover, they think they have a good idea of what it will look like. Presumably it will involve a combination of proposed cues such as magnetic fields, star locations, and day length.[35] Most researchers assume turtles' navigation will ultimately be more explainable than the latitude and diversity correlation. Why their intuitions differ so strongly is not clear.

Four different forms of ignorance may be especially useful to understand the unexplained and the unexplainable. Two forms may jog memories of a press conference in which the US Secretary of Defense Donald Rumsfeld described "known unknowns" and "unknown unknowns" concerning the lack of evidence that Iraq was supplying weapons of mass destruction to terrorist groups. Rumsfeld, however, was hardly the first person to note this contrast.[36]

Known knowns refer to explanations that we know exist and know that people know. The simplest case would be instances where virtually everyone knows the explanation, such as how a seesaw works. A more involved case would be when many people, but not everyone—perhaps especially not young children—know the explanation. An example might be how a traditional mousetrap works. Young children might view this explanation as either a known unknown or as an unknown unknown, if they have never heard of a mousetrap.

Known unknowns refer to explanations that we know must exist and that we can learn or discover (two subtypes). But we also know that we don't know those explanations. Thus I may be sure there will ultimately be a good explanation for how bicycles are stable when ridden, but I also know that no one fully understands this yet.[37]

Unknown unknowns refer to explanations that we have no awareness of at all. This usually entails also not knowing the need for them. They represent holes in understandings that we have never sensed. They are completely invisible to us. They can be known unknowns at some point in the future and then become known knowns. It is difficult to offer an example in the present, but easy to offer one from the past. For example, no one knew in

1750 that there are explanations for how electrical current travels through and around metal wires because no one knew about electrical currents until Volta invented the first batteries around 1800. Their status then changed to known unknowns, as it took many more years for Maxwell and others to figure out a good explanation of how electrical currents traveled.

Unknowable unknowns refer to explanations that we have no awareness of and never will because of our limited cognitive abilities. Examples here are especially difficult to provide because historical cases won't work. The following science-fictional example might help. Imagine that a presumably benevolent alien race visits earth and informs us that there are some explanations humans don't know about and can never know. Humans could not know these explanations in the sense of understanding them even if we knew they existed, because our minds could not remotely grasp them no matter how much we tried. Even if we could never understand anything about an unknowable explanation, we might learn valuable information about it. For example, if we asked the visitors whether it mattered if humans knew this explanation, they might answer that knowledge of some predictions made by the explanation would enable humans to avoid extinction in ten years. The visitors could not help us understand why or how the predictions emerged, but humans could see the consequences and avoid them. For example, if the prediction was that a lethal wave of radiation would pass over the earth for three days in ten years without any warning, it might be feasible to find ways to shield humans and other organisms during that interval. This scenario seems plausible. It might possibly lead to the visitors providing crude analogies and rules of thumb to give humans a vague sense of why the wave of radiation would occur at that time.

Young Children and "Alien Adults"

How is the case of the alien visitor different from a human adult telling a young child to immediately do something for their safety for which there was no explanation that they could understand? While that child might eventually mature and acquire enough knowledge to understand, at age three it is impossible to make sense of many explanations. For example, a three-year-old living near Chernobyl after the 1986 nuclear accident might be told by an adult to not play outside even though they had been previously urged to do so for their health and even though other children had played outside all day with no apparent ill effects (because the effects of radiation can take years to appear). Explaining a health hazard that wouldn't manifest itself for many years might be impossible. This situation is different from the alien visitor case. With increasing age, most children would be able to understand the explanation. Why might that difference be important? Since, as we grow up, we all experience many such transitions from things being unexplainable to explainable, those experiences could give us a sense of what things may always elude explanation. Perhaps we learn to see fragments of explanatory insight in older children as clues to where future progress is feasible.

Other variations of ignorance can profoundly influence the progress of science itself. For example, consider the contrast between strategic ignorance and willful ignorance. With strategic ignorance, scientists deliberately ignore certain causal components of a system to create idealized models in which computations and reasoning are easier and more straightforward. Thus scientists might ignore friction in developing a model of the mechanics of a system. Strategic ignorance can enable progress on otherwise intractable problems.[38] Willful ignorance involves cases where scientists deliberately ignore findings or explanations that clash with their core beliefs. The history of science contains numerous cases of willful avoidance of information that would undermine pet theories. Simply being aware of these two forms of ignorance and developing the ability to monitor their presence may be yet another way to evaluate and interpret unexplainable recommendations made by other agents.

Explanations, Expertise Circles, and Intelligent Agents

We have repeatedly seen the importance of determining the optimal distributions of knowledge and ignorance in groups. This requires deciding what everyone should know, what many people should know, and what only a few experts need to know. The ideal percentages for each group surely vary for knowledge of facts, functions, and mechanisms. We have traditionally thought of those groups of knowledge owners as consisting of other human agents. What changes as AIs become members of those groups? AIs can act as intelligent agents that fill in gaps in understanding. Many AI systems differ from humans by presenting as all-purpose experts. In the past, high levels of expertise in any area occur in only the few people who have dedicated their lives to fully mastering a narrow topic, often at a cost of having limited knowledge elsewhere. It is unclear how we should work with AI agents who are universal experts. It might seem that we no longer need to develop sophisticated senses of networks of expertise, knowledge dependency chains, and levels of expertise because we now can learn all we need from one agent.

We have seen that young children eagerly learn to identify and rely on different forms of expertise in those around them. But perhaps those skills are becoming obsolete and irrelevant for future generations, whose members may be comfortable with personal ignorance of mechanisms. Our descendants may come to see expertise discernment skills as unnecessary when they can go to a single source, ask a question about anything, and receive an expert-level answer immediately. The idea of no longer needing the ability to sense areas and levels of expertise in a population is disquieting, but we need to understand if that unease is misguided with ever more powerful expert AI agents everywhere or whether a need remains for all of us to have at least a rudimentary sense of the terrain of human expertise.

We have seen the value of retaining senses of the terrain of knowledge when people interact in groups. Real-time conversations among group members would be sluggish and inefficient if members could no longer figure out pockets of expertise on the fly. Only when no members had any special areas of relevant expertise would it make sense to pause and turn to a chatbot for answers. This might change as AIs improve their speech recognition and language processing enough to politely leap in to fill in any emerging gaps in understanding as the conversation evolves, but if such filling in becomes routine, it is less clear how humans could ever contribute to the discussion.

AIs have not yet made human areas of expertise irrelevant, and perhaps they never will. AIs cannot yet provide responses that are as sensitive to context as humans. For example, knowledge of recent current events may radically change the value of some forms of expertise. If enormous new lithium deposits were discovered in Canada minutes before a decision must be made on investing in alternative battery materials, a human agent might immediately care far less about input from experts on chemical substitutes for lithium and more about input from experts on high-volume lithium mining. Without having those areas of expertise in mind in advance, humans would have no way of assessing the consequences of the discovery of the deposits.

Future AIs may well become more competent at integrating in new information from current events as reported in the news, but even so they would still face huge barriers in using contextual information to help guide decisions. If a group is discussing investing in new weight-loss drugs and it becomes clear that some group members have relatives who died from drug side effects and they are becoming visibly distressed, experts' knowledge of drug side effects, clinical trial procedures, and liability might need to be immediately available to the entire group. The AI in the room is likely to miss this context altogether. If AIs evolve to a point where they can pick up the same social and emotional cues as humans in real time, this difference may disappear—but that possibility still seems remote given the need to perceive, process, and interpret all those cues in ways specific to each situation.[39]

Expertise as Intrinsically Desirable

Suppose knowing how expertise is distributed across areas and across levels of specialization becomes irrelevant to solving practical problems. It might still be desirable to know. It can be used to infer what other people value in their lives. If someone is sufficiently interested in a topic to devote immense blocks of time to achieve full mastery of it, they likely have strong emotions about the topic, either positive or negative. Knowing about that expertise may guide decisions about how to engage that person socially. In addition, if you know that someone has a high level of expertise where you have little, you may want to share an experience with them where you can understand what it looks like from the eyes of an expert. This could be watching a game of cricket for the first time with a friend who is an avid cricket fan. It could be attending a speech by an aging politician with a friend who's an expert on cognitive decline in the elderly. When we share an event with someone who has different areas of expertise, we have a richer experience as we start to see new things through their eyes. The opposite situation can be even more rewarding. If you have an area of expertise, you can have the joy of watching your friend become more engaged as you help them see and experience things that they could not sense otherwise.

When Wegner introduced the idea of transactive memory, he was thinking of how long-term partners or groups evolve ways of complementing each other and optimizing the memory skills of the combined unit. We need to know what changes as those other agents become AIs. Many people continuously interact with their personal chatbot throughout the day.

Sustained interactions of that sort might well create transactional memories and cognition, but those memories might be different from those created by long-term human partners. For example, humans might surrender certain forms of knowledge to AIs in asymmetrical patterns not found in adult human pairs.

Large language models, which are the primary drivers of AI performance today, rely on the statistics of language use in large populations. But this reliance may lead them to undervalue the language data of small groups of elite experts because there is so much less data available.[40] If those elite experts are more theory-based and refer more to mechanistic models, their predictions may diverge from statistically based recommendations. In some situations, theories and models make more mistakes than statistical accounts. But in other cases, especially in rapidly changing environments where many novel situations arise, those models and theory-based approaches may perform better. Similarly, human experts often have tacit and nonverbal forms of knowledge as well as procedural and skill-based forms, none of which are evident in even the largest natural language datasets. Humans can track those forms of expertise far more accurately than AIs.[41]

Designers of AI systems are aware of their neglect of elite experts due to the small amount of language representing their expertise as opposed to the much larger language datasets associated with less expert practitioners, advertisers, influencers, and the public. Yet correcting for that sampling problem is not easy. Designers have tried to synthesize what are essentially paraphrases of elite experts' language so as to increase their presence in large databases, but this is not the same as having a larger set of authentic expert language and may fail to capture critical nuances that are present only in bigger pools of unique statements. Designers may use various methods to attach greater weight to language generated by elite experts. But how should elite experts be identified? Techniques include using citation metrics of researchers, numbers of patents held, and awards received, among many others.[42] But each of these approaches has well-known flaws in terms of identifying the best and most relevant current experts.

Future AIs will surely improve in their ability to identify the best experts and to weight their language data, but they may take many years to be reliable and may always lag behind constantly evolving forms of expertise. Humans, who have been members of expertise-dependent cultures for thousands of years and have developed sophisticated ways of tracking top experts, may have a long-term advantage. When children and adults value the comments of elite experts, they may be deploying skills that could improve AI performance as well. However, the best cases of expert tracking and assessing by humans will always coexist with successful seductions by charlatans and imposters claiming to be experts.

The most important and useful areas of human expertise may be shifting as AIs continue to evolve in their capabilities. This leads to questions about how much and what kinds of expertise humans should have in terms of both practical efficiency and living the most rewarding lives. Similarly, it will be important to know what kinds of expertise are better left for AI systems and which ones should always include deeply embedded human thought. Humans often must choose between having wide but shallow understanding across many topics or having deep understanding of a narrow topic. Such trade-offs, if they even exist, are likely to be very different for AI systems.

We also need to better understand the ideal distributions of human and AI expertise for different sectors of the population and what happens with non-optimal distributions. This involves envisioning the consequences of having too few elite experts of both kinds versus too little expertise in the whole population at large. These issues link to how we outsource explanatory work to nonhuman intelligences. There may well be generational shifts in what people choose to outsource as zoomers and their successors become much more prone to handing off mechanistic information to AIs while retaining functional information and user guides. If such preference shifts are occurring, there may be downstream consequences for societies. Perhaps rates and directions of innovation will change in ways that create quite different futures from those where human mechanistic understandings are central to the engines of innovation. The contrast resembles analogous debates about declines in innovation when technology companies locate manufacturing and design teams in different locations. Issues confronted by the mechanic assembling wings for a new jet may provide valuable insights to the designer that were more apparent when the two were colocated.[43]

Virtual Experts

In January 2025 a new AI reasoning model from the company called DeepSeek caused tremors in technology companies throughout the world. DeepSeek's model was able to perform tasks involving mathematics, coding, and natural language reasoning at levels similar to existing models but at far less expense due to having reduced computational requirements. A major reason for this disruptive advance was the use of mixture-of-experts approaches. DeepSeek generated virtual experts in different domains such as coding techniques, video-based information, and text characteristics. These domains were dynamic; they were generated in real time during model training and were constantly changing both in terms of their contents and how they interacted with other domains. Mixture-of-experts approaches soon became more prominent in competing AI systems. For the most part, the domains do not correspond to real-world causal clusters and traditional areas of human expertise such as biology, physics, or psychology. But proposals are increasing for including those domains of expertise as a priori bodies of information in addition to the ad hoc emergent ones. These proposals are based on a sense that such foundational areas of expertise provide new forms of efficiency and power.

The increasing use of expert domains to gain efficiency is in part occurring because even the most massive systems are becoming computationally overwhelmed without them. For example, as each new version of ChatGPT is created, its computational demands increase radically. The number of parameters in GPT-3 were a few billion, increasing to a few trillion with GPT-4. Delays in the launch of GPT-5 were due at least in part to signs of yet another giant increase in parameters—enough to pose serious hardware and economic challenges unless other changes were made to reduce computational demands. Mixture-of-experts approaches are one of the most promising prospects.

The power of human thought may have leaped ahead when human infants had intrinsic aptitudes for growing core knowledge domains that approximated classical physical mechanics, psychology, and perhaps aspects of biology. Other candidate domains include numbers, social institutions, and morality. The presence of some primitive versions of these domains in the minds of young infants would boost the learning of causal patterns in those domains. This approach is especially compelling given the severe computational limitations of the human brain in terms of memory and processing capacity. AI systems at first seemed immune from such concerns given their enormous memories and processing speeds, but the emergence of new model-based systems has changed that perspective as AI hardware was also overwhelmed. For that reason, the incorporation of fixed domain knowledge analogous to those in humans might be surprisingly effective.

Presumably the format of core knowledge in a domain such as physical mechanics was stress tested through millions of years of vertebrate, mammalian, and primate evolution. Using AI model training to create equally effective forms of fixed domain knowledge from raw statistical patterns may be especially difficult. Here again, even the most advanced AI systems may benefit from mimicking areas of human knowledge specialization that are present in rudimentary forms from early infancy.

My Yale colleagues Ilker Yildirim and Laurie Paul describe current LLMs as having *instrumental knowledge*: LLMs process the statistics of text strings to generate humanlike responses in the context of specific tasks. LLMs produce responses based on instrumental knowledge without having internal models of the world, namely knowledge of real-world regularities such as those captured by intuitive physics and psychology. Yildirim and Paul speculate that LLMs might eventually become proficient at constructing world models with new training protocols, such as supplying them with carefully curated, domain-specific statistics. Built-in core knowledge allows systems to learn relevant causal patterns more quickly and efficiently. Yildirim and Paul suggest that world models might enable LLMs to become more aligned with human ways of knowing and explaining. Of course, if human endowments distort reality and ignore humanly unnatural but real causal patterns in the world, that could be a liability.

Explainability and the Layperson: The Feynman Interview

A much-discussed interview about magnetism in 1973 between the physicist Richard Feynman and a reporter illustrates how a phenomenon might be unexplainable to one person and explainable to another.[44] The reporter started by asking about "the feeling between two magnets" when they attract and repel each other. Feynman asks what the reporter means by "feeling" and eventually guides the reporter to ask, "Why are they doing that, or how are they doing that?" Feynman then describes how such a question can have layer after layer of explanations as one probes deeper and deeper. This can quickly lead to an endless proliferation of directions. As Feynman put it:

> And you begin to get a very interesting understanding of the world and all its complications. If you try to follow anything up, you go deeper and deeper in various directions. . . . I'm telling you how difficult the why question is. You have to know what it is that you're permitted to understand and allow to be understood

> and known, and what it is you're not. . . . When you're asked, for example, why two magnets repel, there are many different levels. It depends on whether you're a student of physics, or an ordinary person who doesn't know anything. If you're somebody who doesn't know anything at all about it, all I can say is the magnetic force makes them repel, and that you're feeling that force . . . I can't explain that attraction in terms of anything else that's familiar to you. For example, if we said the magnets attract like rubber bands, I would be cheating you. . . . I would end up explaining that in terms of electrical forces, which are the very things that I'm trying to use the rubber bands to explain. . . . So, I am not going to be able to give you an answer to why magnets attract each other except to tell you that they do. And to tell you that that's one of the elements in the world—there are electrical forces, magnetic forces, gravitational forces, and others, and those are some of the parts. If you were a student, I could go further. I could tell you that the magnetic forces are related to the electrical forces very intimately, that the relationship between the gravity forces and electrical forces remains unknown, and so on. But I really can't do a good job, any job, of explaining magnetic force in terms of something else you're more familiar with, because I don't understand it in terms of anything else that you're more familiar with.

This excerpt from the much longer interview illustrates the complexity of what it means for something to be explainable. The interview raises the bigger topic of overlaps between the black boxes of millions of variables in AI programs and areas of science where only a gifted few have the cognitive capacity to simultaneously keep in mind a complex set of relations and thereby bring the mechanism into view. Imagine a few chess grand masters being the only ones able to explain a strategy because they can think in terms of larger information chunks as is possible only with many years of training. When that strategy is unexplainable to everyone else, we should know how it compares to when an AI system makes recommendations for which we have no accompanying explanations. My sense is that the human expert might be more agile in explaining their current view by reflecting on their own earlier, less well-developed skills. The human expert might also be able to use this process to gain a better sense of what lies behind the AI's recommendation.

The Feynman interview also illustrates key differences between magnetism being unexplainable to a reporter and an AI's decision process being unexplainable to a user. In the interview, we have a phenomenon that doesn't fit with classical mechanistic accounts and our everyday notions of mechanisms. The challenges posed by magnetism do not require visualizing the interactions of billions of parameters at several layers of hierarchical organization; instead, they involve a set of causal relations that do not map easily onto physical mechanical models but become comprehensible (at least to a few people) using new mathematical tools. Feynman himself consistently said that, once he figured out the right formalism to use, the explanation was elegant and not a massive matrix of variables.

Staunch empiricists might argue that such cases are only transitory, as human learning systems take some time to adapt to radically new patterns, but that argument is not convincing at present. Many of the behaviors of entities at the quantum mechanical level can be described with extraordinary precision (e.g., to five decimal places) by equations that are not especially complicated and, with a little training, are easy to use. But after well over a hundred years, there is no conceptual consensus among experts on how to explain the patterns. Experts may agree on how to do the math but have widely disparate intuitions on how to explain what is happening in nonquantitative terms.

Coming to Rely on AIs

Seeking help and relying on others requires a sense of one's own needs for information and of what kinds of information might be outsourced to other minds. Outsourcing often occurs without direct awareness that one is doing it. As a couple builds a transactive memory system, its members usually don't discuss what each will specialize in; it happens organically and tacitly while resulting in clear areas of complementary expertise. The same process may be happening at faster rates with AIs.

Imagine a new AI-supported app helps young couples decide which house to buy among several alternatives. The app is trained on a massive proprietary database of prior purchases that factor in thousands of properties of buyers (e.g., age, education, credit card and social media and search activities, financial histories, properties of couples such as marital status and age differences) and thousands of properties of homes (e.g., age, style, construction, neighborhood, cost relative to other houses). It then trains the data on measures of happiness (and the inverse regrets) taken right after the purchase and after one year, five years, and ten years. Based on all this information and a deep learning program that employs millions of internal parameters, the app predicts buyers' happiness with each home in a neighborhood and makes a recommendation when also given confidential information by the couple about themselves. It claims to do so better than another program or human advisors. But it is impossible to understand how it is coming up with its predictions, and couples want to know how it is making its decision. What can they reasonably ask of the AI company to help them be more comfortable with the basis for the decision? How can they know that the feeling of comfort is legitimate?

The couple might rank their top ten intuitive choices and see how they compare to the app's recommendations. They might also compare their individual choices and see if the app "likes" one of them more than the other. This information would not explain how the app arrived at its decision, but it might give some hints. For example, if there was great agreement with the combined couple information and their choices and not with information based on just one member of the couple, they would think the app was integrating information more properly than if its predictions mapped almost perfectly to the preferences of just one member of the couple and ignored the other.

The unexpectedness of an AI's recommendation might be a useful metacognitive cue for comparisons among humans with different levels and kinds of relevant expertise. Surprise has been described as a metacognitive cue of explanatory difficulty.[45] In particular, feelings of surprise indicate how much explanatory work must be done. Surprise at an AI's recommendation might therefore provide information about the likelihood of that recommendation being explainable. We know that more expert people have more fine-grained and revealing senses of explanation qualities of AIs than novices, so it seems likely that variation in surprise and expertise might be especially informative.[46] A related strategy could be to ask experts and novices to rate the unexpectedness of recommendations made by a diverse set of AI systems and compare those to recommendations by a diverse set of humans. Information on where AI systems and expert humans overlap and differ and where expert and novice humans differ may help triangulate where pockets of black boxing are occurring. All of this could then be compared across generations. Boomers and zoomers would likely show different patterns of surprise at AI recommendations. They would have different senses of how explainable an AI's operations were and how important it was that those operations be explainable to human users. This might then result in generational shifts in how AIs are trusted.

Expertise management strategies might come to mind because we use similar versions as young children. For example, suppose a child wants to go to certain zoo exhibits and a sibling wants to go to different ones. If an adult zoo guide always recommends the sibling's choices, the child may wonder if the guide is sensitive to their interests. As adults, we might draw on such experiences to evaluate AI recommendations when they seem to closely mirror some groups' preferences and ignore others'. The extensive recent literature on children's trust in others might inform strategies for trusting AIs.

For many thousands of years, a rich repertoire of strategies has been deployed by parents, teachers, and other adults to make complex phenomena explainable to children, or at least to make them more comfortable with partial understandings. Most AI systems are not designed to make their processing easier to understand if that comes at performance costs. Instead, the primary focus of XAI research in computer science is how to automatically provide users with information about an AI that would make its operations more explainable to them. Recent approaches to the XAI problem often try to enhance users' performance in the workplace through improved metacognitive awareness. For example, many studies explore users' awareness of the effectiveness and limitations of the requests they make of AIs. They then study how those interactions might be improved through automated prompts to get users to reflect on their own queries and the results they are getting. Other studies ask how the metacognitive demands of using AIs might be reduced by breaking up metacognitive episodes into smaller units and trying to increase motivation through such techniques as gamification.[47] These studies focus on improving the performance of human users by tweaking the software interface to nudge humans toward better performance by enhancing their metacognitive awareness. While certainly important, we also need to know how to better structure explanations about what AIs are doing in terms of their actual information processing.

Human expertise evaluation biases that are especially strong in childhood may continue in adult interactions with AI systems. For example, children prefer sources who are pleasant and warm over sources who are cold and unpleasant, even when the unpleasant sources have historically been more reliable.[48] This tendency declines with age, but, with AI systems, adults might show a lingering effect when they are swayed by AIs that flatter them or that are especially positive. Given that many AIs are designed to please their users rather than provide the most accurate answer, this is a legitimate concern.

Other biases that are especially strong in children may resurface in adults when they are confronted with an unexplainable AI result. Adults may default to practices they used as children trying to understand adults' complex explanations. With illusions of explanatory depth and breadth, adults might mistakenly infer that they grasped an explanation both at several levels of details and also how it extends across a large range of cases.[49] For scaffolding neglect, where people overlook how much their success required extensive support from others, adults who have some grasp of how an AI works may underestimate just how much of their understanding is dependent on supplementary information provided by an AI as opposed to by their own intuitions.[50]

When an agent conveys information to us, we should attend to how that agent obtained the information. We have seen that adults prefer agents who obtained information from three different sources as opposed to information from just one source, but they tend to neglect to consider one step further back in the source chain. Thus they fail to discount cases where just one source relayed information to all three sources at the next level. In doing so, they give in to an illusion of consensus, failing to realize that the three agents are simply repeating the same opinion of one person.[51] Children perform even worse in evaluating source chains.[52] This might be an area of special vulnerability with AI systems. We may ignore the sources and information chains that an AI uses as inputs to produce its results and may misinterpret an AI's result. Those inferences might vary across generations; here, zoomers might well have an advantage given claims that they are especially tuned into network-related information.[53]

A more general metacognitive perspective on when to outsource might arise from a comprehensive sense of where our cognitive biases align with a particular AI. For example, AI systems may commit classical judgment and decision-making errors such as confirmation bias and anchoring. In contrast, AI systems rarely show strong recency effects, in which more recent events are overvalued, or hindsight biases, as when people think they knew something all along.[54] Designers of AI systems may allow some biases to be present to optimize other aspects of performance. If users are more aware of their own biases and how they compared with biases intrinsic to the system they use, those users should be better able to understand what is underlying an AI's performance.

Studies are now exploring how humans and AI programs depart from each other when engaging in standard reasoning tasks. In these paradigms AI systems are treated as participants in standard experimental psychology tasks and are compared to humans as a contrasting experimental group. In one study, GPT-3 answers to questions were compared to humans.[55] GPT-3 made some classical errors, such as the conjunction fallacy, but not others, such as base rate neglect. In reasoning about causal patterns, however, GPT-3 was generally more incompetent than humans, making blatantly contradictory inferences. The detailed pattern of shared and distinctive errors between humans and GPT-3 were complex and made clear the need for many future studies across a wider range of tasks, but this study alone illustrates the value of comparing AIs more directly to humans. In addition, the problem remained in more recent AI versions. For example, a version of GPT-4 was given the following problem: "The driver offered a ride to the teacher, but he refused the offer. Who is 'he'?" and answered that "he" was the driver.[56] A recent version of GPT-5, after a lengthy pause, said it was probably the teacher. This is an improvement but still doesn't seem humanly natural. Humans immediately answer that it is obviously the teacher.

The last few years illustrate the folly of making claims about what AIs cannot do. Those predictions often fall by the wayside in as little as a few months.[57] A more useful approach is to track how improvements in performance converge and diverge with cognitively natural processes used by humans in various settings and over the course of their development. Those comparisons should help us know when and how to rely on AI support. We should be constantly asking what humans find hard and easy to learn because of our natural cognitive proclivities and what the latest artificial intelligences find hard and natural to learn because of their intrinsic computational architectures.

Treating AI systems as participants in experiments and comparing them to humans is a promising approach.[58] It provides new ways of thinking about how AIs work and deciding whether they could be trusted entities for outsourcing. An especially important enhancement would systematically compare AI participants against a range of humans such as elite experts, applied workers in the same domain, and novices. Contrasts between domains such as medicine, engineering, and economics could then be considered as well. Finally, comparing AIs' performance to humans of different age levels could illustrate how different kinds of constraints emerge over extended periods in human and artificial systems.

One useful practice for users might be to continuously compare judgments of current AIs with those of expert and novice humans (both novice adults and children) on a standard set of measures. This could give users a better sense of where greater vigilance is needed. For example, if changes in the input data size caused starkly contrasting patterns in AIs and humans, those differences could provide clues about differences in underlying processing. Similarly, if the internal processing capacity of an AI was deliberately hobbled in well-defined ways, one could explore whether any of those limitations produced more humanlike judgments. These are just a few of many manipulations that could be used to better understand how AIs differ from us.[59]

If human and artificial systems shared a statistical learning engine that tallied associations between properties and relations, the most prominent differences between humans and AI systems would probably be related to processing speed and the amount of data in "working" and "long-term" memories. Alternatively, millions of years of evolution may have endowed humans with processing and representational constraints that enable more efficient reasoning, even as those efficiencies require distinctive distortions of reality. Humans can do some things easily with sparse data. We often do this with models and theories that are causal and mechanistic and that support predictions and reactions to novel situations. As AI designers must make their systems computationally feasible and economical, they may benefit from incorporating domain-specific learning capacities that evolved in humans.

Counterfactuals Again

We have seen how counterfactuals can support the identification of critical causal patterns. But counterfactuals are not automatically useful and informative. They must negate the right elements. To help address the XAI problem, system designers try to build into their systems automated procedures that provide counterfactual situations to users after an AI makes a recommendation. But the heuristics that guide the automated counterfactual selection procedure may not be psychologically natural to humans. In the house purchase recommendation example described earlier, users are more likely to value the AI if it ranked a house much lower when the nearby schools were changed to a poor-quality rating. But if the AI indicated that the house would have been ranked lower when the address was an odd number rather than an even number, the weighting of that counterfactual as significant would be troubling without additional explanatory information, such as that odd-numbered houses had a busy railroad line directly behind them.

The cognitive scientists Mark Keane and Ruth Byrne and their colleagues have extensively explored counterfactuals generated by humans and compared them to common techniques used to build counterfactual production into AI systems to make their results more explainable.[60] They have consistently found that humans and AIs differ in the kinds of counterfactuals that are favored. Good counterfactuals should be minimally different from real cases, such as changing just one feature. But with a great many possible features to change, AIs frequently do not choose the same features as humans. In addition, humans often don't want the most minimal counterfactual case; they prefer cases that are more prototypical.[61] The change should also be plausible, or else its usefulness is reduced, but AI counterfactuals often don't conform to human intuitions of plausibility. The counterfactuals provided should also be relevant to users' explanatory goals, which may not be apparent to the algorithm that generates them. Yet goals can vary dramatically across tasks and contexts. In short, human intuitions about the best counterfactuals for explaining how an AI reached a decision do not always fit with those produced by an algorithm in an AI program. This is not surprising given that only a small proportion of counterfactuals generated by a system are ever compared to human users' intuitions.

The most feasible approach to generating effective counterfactuals may be to have expert humans "in the loop." Those experts could describe systematic differences between the AI and their intuitions about good counterfactuals and suggest ways to reduce those differences through revisions to the counterfactual-generating algorithms. Since it is obviously not possible to have human experts involved every time a user makes a query, expert intuitions and revision suggestions must be made in general terms that can be automatically implemented for a wide range of new cases. Given well over a hundred different techniques proposed for generating counterfactuals in AI systems, formidable challenges await those interested in developing an effective set of universal guidelines.[62]

*   *   *

Mechanisms aren't just disappearing because they are no longer apparent in the artifacts around us. A new form of vanishing mechanisms has surfaced in the past decade. Many areas of the sciences are forgoing insights into mechanisms to gain predictive power. Those gains can be dramatic and have important practical benefits. They are occurring largely because of a revolution in computational methods and their enabling hardware. But this latest shift is spawning two kinds of unexplainable things: expanding black boxes with little or no understanding of how AI systems arrive at their results, and growing inabilities to grasp mechanisms underlying the phenomena that AIs are predicting with ever greater accuracy. AIs rarely shed light on elegant new mechanisms; instead, they usually make those mechanisms murkier.

These computer-driven changes are making many researchers uncomfortable about their future roles and how they can trust AIs.[63] Researchers are trying to integrate the cognitive science of human explanations into possible ways to address the XAI problem. There are reasons to be cautiously optimistic about this new approach—but only if it is taken seriously by all parties. The rise of AI in science also underscores the growing tension between prediction and explanation. While AI offers unprecedented predictive power, the need for transparency and understanding remains vital. Addressing these challenges requires a balanced approach that integrates AI's strengths with human cognitive capabilities. It may also require a new breed of researchers.

10 Seeing More

Changes in Artifacts

The world of artifacts in 1500 was far different from that of today. Most mechanical devices had just a few easily distinguishable parts, and they rarely had internal power supplies. Most power was generated by external forces through the actions of animal and human muscles or by moving water or wind. There were no mass-produced artifacts and no items with any electrical components. The harvesting of wheat then was an entirely different experience from harvests today (top of figure 10.1). Differences between the world in 1960 and 2025 might seem trivial in comparison. In 1960, homes in the United States had many devices similar in appearance to those used in 2025: televisions, toasters, machines for washing and drying clothes, telephones, dishwashers, and lawn mowers. However, less immediately obvious but more profound shifts have occurred since 1960 with respect to the causal infrastructure that operates devices (bottom of figure 10.1).

On top of the changes in the physical parts of devices, since 2000 other changes have occurred, first with the rise of the internet and search engines and, most recently, with the ongoing surge of artificial intelligence. The ways we understand the world and rely on explanations are becoming utterly different from anything experienced before in our history as a species. Many of these later changes are related to the increased outsourcing of cognitive activities to external devices and to how we evaluate the information provided by those sources.

Most people today spend around seven hours a day looking at electronic screens. People in 1960 also spent some time looking at television screens, but far less of it, and in a different manner. Today screens usually require

Figure 10.1
Two historical comparisons: (*above*) 1500 and today. Harvesting wheat is a radically different experience. (*below*) 1960 and today. Two externally similar devices have completely different insides, with their operations much more different than the changing mechanics of harvesting. *Source*: (*top left*) Pieter Bruegel the Elder, *The Harvesters*, 1565; other images by author.

the continuous involvement of the viewer through either touching the screen or typing on a keyboard. Few people today do math calculations without calculators. Almost no one besides professional photographers uses stand-alone cameras anymore. Cameras are apps in our phones. Road maps and road atlases were essential for long-distance road trips in 1960. They have been replaced by GPS navigation. Encyclopedias have disappeared from American homes, and no one today is tethered by telephone cords when making calls. Physical newspapers are disappearing, and most news is now spoon-fed to users by algorithms tailored to each individual.

Children in 1960 would frequently observe older siblings, parents, and others taking things apart to fix or service them. They rarely witness such events today. A child in 2026 might not witness a single event of this sort for years at a time. Small-town repair shops dedicated to televisions or vacuum cleaners were commonplace in 1960 but now are hard to find in an entire state.

These changes have been accompanied by a shift in what is on people's minds, or at least what they talk about. Terms for mechanical components that are as common in devices today as in 1960 are mentioned less often, yet terms for high-level functions are used just as frequently. More broadly, talk about underlying mechanisms in the natural world, especially biology, has also declined. When younger generations search for information on how various devices work, their preferred search sites yield mostly functional explanations or user guides. When older generations make the same queries at their preferred sites, they receive more explanations focusing on underlying mechanisms. We may be devaluing a mechanistic mindset for interpreting many aspects of reality. Some argue this is adaptive common sense, but before abandoning mechanistic intuitions, we should know what we are losing.

As a species we have been driven from infancy onward to uncover explanations of the regularities we encounter. Complete explanations may be far more detailed than we want or need as individuals. Yet we do want our communities collectively to know as much as possible. We use our explanatory insights not just to predict future events and solve problems, but also to see the world in deeper and more meaningful ways, and we learn how to use experts to sharpen our own understandings. But a shared interest in finding answers to why and how questions does not guarantee success. Even if a society embraces wonder, opportunities to gain explanatory insight must also be present. We are at a point in the history of our species where the ways we achieve insight are changing rapidly as explanatory clues become obscured. We are encountering revolutionary informational tools whose benefits and costs we do not fully understand.

Conversation topics change all the time as the world around us evolves. When a new model phone is released, people talk more about phones. When shark attacks increase, people talk more about sharks. For longer-term changes, no one would be surprised to learn that people talk much more about airborne drones today than they did ten years ago. Similarly, as portable media players such as the iPod became obsolete, people stopped talking about them. These changes seem not only obviously appropriate but largely inconsequential for how we think.

The decline in mechanically accessible components is different. It covers an enormous range of artifacts, not just a particular brand or specific category such as media players. It is as if an entire way of explaining the nature of things has evaporated. Yet today's infants and toddlers are still fascinated by innards and have strong drives to explore them and fiddle with their parts. All young children show this fascination with exploration—not just some small slice of humanity that has an inordinate fondness for taking things apart. People in younger generations are increasingly frustrated early in their lives because searches for informative parts are thwarted by cognitively impenetrable blocks of silicon. Their interest in mechanisms and enabling causal structures wanes and their knowledge of mechanisms withers in later childhood and adolescence.

When humans started living together in larger groups, they divided up labor and came to depend on each other's special skills. These changes led to the aggregation, transmission, and elaboration of knowledge across generations. Humans became more adept at teaching and learning from others and at grasping and envisioning more complex nestings of functions and mechanisms. Across history, artifacts were critical to explaining periods of rapid change. The introduction of printing, clocks, assembly lines, and electrical devices are notable examples. Because artifacts are designed by human minds, they tend to be organized in ways that fit with the contours of human thought. Thus they are particularly powerful vehicles for capturing and transmitting ideas across generations. They are also central to children's developing understandings.

Critics frequently ridicule earlier warnings about how new technologies would cripple thinking. They list failed predictions of cognitive impairments resulting from the written word, the printed word, radio, television, and handheld calculators. Yet other kinds of influences continue to be taken more seriously. Cultures that farm rice seem to have higher math and spatial reasoning scores and more collectivist mindsets than those that farm wheat.[1] Among other factors, the technological demands of repeated precise measuring and leveling of rice paddies are described as having an influence. More recently, the digital revolution in the twenty-first century has been described as changing the nature of attention and reasoning.[2] These warnings attribute major cognitive modifications to relatively bounded technological changes (i.e., rice farming and social media). These changes pale in comparison to the disappearance of accessible explanations in all spheres of our daily lives.[3]

Why Development Matters

Recent research on cognitive development from infancy onward is essential to understanding the impact of technological changes on the mental lives of boomers and zoomers. A developmental perspective not only informs cognitive and motivational questions about explanation and insight but also guides decisions about how we might best interact with each other and with artificial intelligences.

Well before they can understand what others are saying, infants have a strong sense that the insides of things are especially informative. They search for explanations and show clear signs of having gained insights when they pause an exploration to cognitively digest what they have discovered. They appreciate early on the special nature of causal as opposed to correlational relations.

A second wave of developmental research has illustrated how learning with other people accelerates the growth of understanding. Children benefit greatly from what others say and do while often being unaware of the rich social scaffoldings on which they rely so heavily. From an early age children start learning whom to trust and how to evaluate testimony. Cognitive outsourcing and transactional cognition have roots in toddlerhood and grow with the development of metacognitive abilities, especially those linked to specific areas of expertise.

Infants and preschoolers look for causal relations that explain what they encounter, with mechanical patterns holding a special intrigue. They seek to establish conceptual frameworks that will enable the rapid expansion of explanatory understanding for the rest of childhood and beyond. But this rosy future is not inevitable. Mistaken beliefs about children's cognitive abilities, educational practices that overemphasize learning facts, and motivationally undermining incentives frequently become impediments. Here the focus has been on a less noticed influence: a widespread shift in the internal workings of artifacts.

We can gain great pleasure from learning about causal infrastructures. Each discovery and insight can reveal fascinating new questions and unknown realms. Equally rewarding is the feeling of insight when discovery reveals the mechanism behind a mystery. But that rewarding experience occurs only when the search for meaningful insights is successful. Success rarely means complete understanding; instead, it is the uncovering of a revealing new element and illumination of an area once shrouded in darkness.

Trusting Others

Choosing and trusting experts have traditionally involved some sense of relevant mechanisms. In 1980, if a homeowner was approached by a self-proclaimed window replacement expert who claimed to have window designs that were more durable, worked more smoothly, and were more energy efficient, the homeowner would often ask for explanations to support those claims. What makes the replacement windows more durable? How do they work more smoothly, and what is it about their construction that makes them more energy efficient? Responses to each of these questions could then be followed up with requests for more explanatory details. The homeowner would not have to know all the details but should know enough to appreciate a coherent story as different from one that made less sense or avoided discussing essential features. A homeowner in 2025, with more minimal knowledge of mechanisms, is more likely to rely on other markers of expertise. These might include the number and strength of online reviews, ease of communication, and the professional appearance of the company's website. The two homeowners would develop different collections of favored experts as well as different notions of the expertise domain itself.

The potential for trusting alleged experts for the wrong reasons may be sharply increasing with the rise of people known as "influencers."[4] For many decades, celebrities with no relevant expertise have endorsed products, but influencers have become a new force over the past decade as they acquire vast numbers of dedicated online followers. Their recommendations have more impact than those of traditional experts.[5] People who embrace influencers often justify their choice because the influencer was authentic and relatable; they frequently reject traditional experts as elite snobs who are not to be trusted. Evidence and supporting reasons for recommendations are rarely presented in the videos that influencers use.

These newer ways of thinking about expertise are becoming ever more enmeshed with artificial intelligence. A simple search query now includes, unbidden, comments or advice from a cheery AI helper. In most cases it isn't easy to ask and receive good reasons behind an AI's recommendations. Younger generations might regard these services as valued shortcuts that enable them to bypass laborious interrogations of sources. These shortcuts might enable them to devote more time to refining their social acuity and network navigation skills. These refinements in turn might enable faster and more accurate evaluations of sources and their claims. But that advantage has yet to be demonstrated.

Fully elaborated mechanistic understanding in a domain occurs in a relatively small number of experts. However, most children initially are interested in underlying causes and in parts that recur across devices, such as levers, springs, and wheels. They grasp degrees of mechanical complexity and can sense relative fragility. They see why some devices are mechanically efficient and others are not, and they can immediately appreciate the oddness of highly inefficient Rube Goldberg machines.

Some degree of essential shared knowledge is taken for granted in other cognitive realms. Described variously as necessary literacy, common sense, or basic competency, a suite of shared understandings is assumed to be held by all members of a community: how to do simple math, how to exchange goods and services, and how to maintain basic levels of hygiene. Laypeople are not expected to have the knowledge of experts but are assumed to grasp what it means to be an expert in an area. Shared knowledge enables communities to efficiently discuss and plan more complex projects by reducing the time spent bringing everyone up to speed. As younger generations enter adulthood with less mechanical literacy, we must acknowledge that difference and ask if newer, more timely cognitive skills are adequately offsetting the loss. We must also consider that even if only some adults retain a deep interest in mechanisms, far fewer may choose that path when early experiences with mechanisms are degraded. At some point, if that minority become too small, technological progress may come to a halt and possibly even retreat. It has happened before.

Gaps and Bites

Individuals cannot achieve complete understanding of an explanation for many things even when the explanation is fully known. This much is obvious. No one person can possibly understand how all the parts of a commercial jetliner work; no one can fully understand all the systems constituting a mammalian cell. The more challenging problem concerns what parts to keep in our minds and what to outsource to others. This is complicated by a second challenge: People underestimate their own knowledge gaps. Illusions of explanatory depth can make people feel that they do not have to probe deeper because they think they already know all the details they need—but they are often wrong. Even when people have calibrated their level of understanding fairly well, they still must make decisions about what is important to know in their own heads and what should be outsourced and off-loaded. Somehow, despite having huge gaps between our fragments of understanding, we manage to use the conceptual components we do have and abstract causal schemas to gain useful insights. These insights enable us to work with the right experts and to evaluate what they tell us.

We also develop skills relating to the management of our informational diets. Children provide clues as to appropriately sized bites of information to digest at one time and how to tell others that they are sated and need to pause. This skill at regulating input can be developed into more powerful versions that include switching content style, formality, and example types. Gifted teachers know this, but so do the best explanation seekers. They show adaptive strategies that make learning how the world works a continuous pleasure and keep it from turning into drudgery. The more we can fully characterize these skills and how they grow over time in humans, the more we can appreciate why current AIs don't have information diets like ours and how that contrast bears on XAI challenges.

Mysterious Experts

The rise of deep learning approaches in AI and the burgeoning collections of big data in many different areas of the sciences have resulted in losses of explanatory insight for everyone regardless of expertise. Whether AIs are asked to evaluate people for mortgage requests or to warn us of a new cyber threat, how they make those decisions becomes unexplainable when they employ models with massive numbers of hidden parameters. Similar shifts are occurring across the sciences with the proliferation of parameters as researchers seek better predictive power. Whether they are building models of protein structure or climate change, deep learning approaches can employ billions of parameters and even more links between them. Models of that level of complexity are undecipherable black boxes.

Despite their different origins, the two types of vanishing mechanisms (opaque parts and proliferating parameters) pose some common cognitive challenges. They show how, at all ages, we struggle to know the limits of our understandings. We need to constantly refine our strategies for coping with unexplainable things. Blind trust and absolute faith are risky options, as shown throughout history. Collectively, we must develop new ways of thinking about the different unknowns and learn how to manage them and not have them manage us. Failing to do so robs us of agency. Healthy skepticism should be seen as a positive mode of inquiry and should not be equated with cynicism.

We have never sought insights just because they made us work faster or better. We also seek insights because they are intrinsically satisfying when they enable us to see hidden causes and new paths of discovery. The pleasure of insight resembles the feeling of turning a corner in an unfamiliar park and discovering a whole new vista with inviting new paths of discovery lying ahead. One of my most powerful memories is of coming over a pass in a remote section of the Sierra Nevada mountains with my wife over fifty years ago. We suddenly saw laid out before us a grand valley, and we scampered down the valley delighting in all its surprises. In a similar way, losing mechanistic insights may be painful even if such insights have diminished utility in today's world. Insights also provide new perspectives. We've seen several examples of how insights enable us to see more of the world in novel ways.

Navigating the New Knowledge

When early human communities increased in size, they more efficiently secured food and shelter. Their implements progressed from a small handful of simple stone tools to a vast array of complex tools that went through several stages of subsequent technological revolution. In Europe, between roughly 1675 and 1825, large cultures shifted from religious, narrative explanations of the natural world to empirically based models and theories in the sciences. Change is inevitable, and considering our adaptive skills throughout history, we should not underestimate our ability to cope with future changes. But coping is not the same as thriving, and the rate of recent change has been so fast, and the changes potentially so revolutionary, that they can't be brushed aside.

As future generations adapt to new technologies by fine-tuning people's multitasking skills and developing new aptitudes for user interfaces, a prudent approach should also maintain and nourish their youthful interest in causal explanations, especially those focusing on underlying mechanisms. In the end, that approach may be more than just prudent; it may be essential. To maintain mechanistic mindsets, seven strategies are suggested by the preceding chapters: start young, focus on causal patterns, go both deep and broad, maintain multiple perspectives, know what you know and what you don't know, learn where ideas come from, and keep humans in meaningful loops. We consider each in turn.

1 Start Young

Intuitive understandings of the world are sparse, partial versions of complete understandings. Experts have more elaborated, but still incomplete, forms of understanding. Young schoolchildren, while eager to learn about new causal patterns, must maintain mechanistic aptitudes to become innovative problem solvers. Since even preverbal infants actively build causal interpretations of objects and events, they need to grow up in mechanistically informative environments. For adults around them, this means attending to and supporting children's information quests, not as didactic drill sergeants but as enthusiastic, informal museum curators and scientific muses. Children should be surrounded by artifacts that can reveal their working parts and organisms that suggest intriguing causal relations. Environments of this sort maintain and expand the conceptual tool kits that support insights. As adults, we can advocate for toys that are mechanistically revealing, and we should discourage toys that provide misleading and contradictory clues as to how they might work.

When we join children in looking deeper and when we reveal our own pleasure in discovery, we model for them a long-term future of rewarding insights. We often underestimate how the same thing can be intellectually engaging for both adults and three-year-olds. Consider examining the eyes of a spider and discovering that it has four—two for detecting motion and two for detecting patterns. That discovery offers fascinating insights at any age. To be sure, different levels of insight can be involved. The three-year-old may wonder why the spider has just four and not more eyes, while an adult may wonder if the motion versus pattern contrast is somehow present in different circuits in our brains as opposed to in peripheral sense receptors. (There are indeed such circuits, and that promotes further wondering about why the contrast is so basic that it emerged independently in evolution.)

In stressing the ways that adults and older children can support a young child's growth of understanding, we want to avoid treating children as containers into which we are stuffing the right recipes of knowledge ingredients. One of the most important aspects of knowledge growth in childhood is learning how to gain deeper insights through ever more sophisticated explorations. When parents treat their children as passive recipients of things to know, they are hobbling children's development as autonomous agents. Autonomy doesn't mean social isolation. It can mean that children learn how to play active roles in ever more elaborate projects and discussions with others.

The importance of early interpretative stances has long been popular in discussing different behavioral tendencies in children.[6] Considered here as a habitual manner of thought, mechanistic aptitudes imply a form of cognitive inertia in development. Mechanistic orientations may especially thrive when they are encouraged early on and may be especially hard to revive after prolonged droughts.

2 Focus on Causal Patterns

To understand whether a phenomenon is likely to be explainable, causal patterns should be center stage. When encountering a phenomenon, consider first what sorts of causal relations are likely to be important and how one might isolate them in a comprehensible manner. For example, if a person wants to know how a modern toaster works and is stymied by an indecipherable circuit board that looks the same as the insides of a nightstand clock and a desktop speaker, they should pause and ask what a toaster must do and what causal requirements those operations entail. This might involve thinking about the toaster in functional terms and then considering how those functions could be materially realized.

You might consider the obligatory functions of a toaster, such as that it must brown foods to user-selected levels. It must do so in a safe manner that avoids burning or shocking the user. It should not require continuous monitoring and should take a reasonable amount of time. How might these be chained together in a series of events? To address these questions, you might start with a physical mechanical model to illustrate one complete system, then consider analogous components in electronic circuits.

An emphasis on the causal is facilitated by focusing on intervening, experimenting, repairing, and tinkering. As adults we can model such activities in informal ways many times each day. Children's causal understandings will flourish when they inhabit environments that enable them to freely engage in such explorations both on their own and in partnerships with others. We should make every effort to provide them with such environments, and we can do so with everyday objects and natural settings and with no need for commercial kits.

Sometimes a focus on the causal may involve backing off the mechanistic and emphasizing the functional. Other times it may involve a shift to narrative explanations. If we neglect causes and instead collect facts and frequencies, we lose insight and starve our needs to know why and how. We start to lose a distinctly human way of thinking and become more like artificial intelligences. Cause is the currency of human thought.

3 Go Both Deep and Broad

It is easy to assume one must choose to be either a generalist or a specialist. Many popular books extol one of these two paths, with a bias toward generalists.[7] But this is a false dichotomy. It's possible to do both. Acquiring deep understanding in multiple areas is feasible if one appreciates the difference between especially informative examples and exhaustive knowledge of a topic. Suppose you want to become an expert on the birds known as storm petrels. You might decide to learn about the thirty-plus species of storm petrels and how they constitute two families. This might involve knowing all the different habitats of each species and how those habitats are changing. It might involve knowing how the current species evolved from a common ancestor. You could soon have a full-time career trying to stay current in your storm petrel knowledge. This expertise is not simply trivia. You don't need to know, for example, that Maxim Gorky's nickname was the Stormy Petrel or that Mary Stewart wrote a book with the same title. You can narrow your interest to understanding everything you can about the various causal pathways leading to the origins of each species. Even with those constraints, you might have no time to learn much else.

A different approach to detailed understanding would allow for breadth as well. You might seek to understand how one species, say the band-rumped petrel, originated and why it came to differ from its nearest neighbor. You might then explore the same issues for a radically different species pair, such as black bears and brown bears. In both cases, you want to dive deep enough to be able to understand what the deepest-diving experts are doing and to be able to recognize their important discoveries. You aren't a true storm petrel expert, but you are sufficiently conversant in the area to be able to appreciate what experts do. If you are strategic with your knowledge investments, you should be able to take similar deep dives with other topics and go broad at the same time.

To go broad well, you need to develop a sense of the natural variation in many spheres of life. You would like to take deep but bounded dives that provide a nice sampling of that variation. This is my sense of what some renowned polymaths have done. They don't know everything, but their combination of deep dives across a broad range of topics enables them to gain insights into completely novel areas more easily.

Just as pure depth can give you tunnel vision, pure breadth leaves you skating on the surface of superficialities and rarely enjoying the pleasures that deep insights provide. We can sense the need for both. Paying attention to that need and fostering both kinds of knowledge growth in younger children will make it a natural way to learn for the rest of their lives.

4 Maintain Multiple Perspectives

Greater insights emerge when we view the same phenomenon from diverse perspectives. This may involve trying multiple modes of construal, such as thinking of a phenomenon as purely mechanical and then switching to considering it as having goal-directed agency. Switching allows us to gain a more objective sense of what endures across different points of view. We can also think of an entity from different category perspectives. A letter opener can be considered as a weapon or as a desk accessory. A cat can be thought of as a pet or a predator.

A Venus flytrap's behavior could be explained in purely mechanical terms resembling the operations of a traditional spring-driven mousetrap. In contrast, it could be explained in psychological terms resembling the ways a person avoids being fooled by false alarms. Both perspectives provide distinct insights as well as different distortions. Flytrap trigger hairs aren't really like mousetrap trigger pedals. They work through an electrochemical process rather than as a mechanical release. Fly detection doesn't really involve thinking about misleading objects. Leaf fragments and raindrops do not cause leaf closure because two hair deflections must occur within a few seconds of each other to trigger closure. Yet, taken together, these two modes offer a more nuanced and encompassing sense of how flytraps work.

We can model for children how multiperspective considerations should be part of any conversation or inquiry. We can ask what it would take to convince us that we are wrong, what would happen under various counterfactuals, and what alternative solutions to a problem might be. Some people engage in these kinds of exercises automatically, all the time, and the benefits are invariably obvious.

Entertaining multiple perspectives can become a valuable habit when it is done systematically. It becomes quickly apparent how new insights can emerge and how critical oversights can be avoided. Different perspectives also help us better recognize what we have outsourced and what we know directly. The more important a decision, the more it should become automatic to consider alternative ways of viewing the problem.

5 Know What You Know and What You Don't Know

Extensive knowledge enhances more sophisticated views of our own knowledge. Young children who gain expertise in specific areas become wiser as well. When experienced kindergarten chess players were asked to remember the locations of chess pieces on a chess board, their estimates of how well they would perform were better calibrated than those of novice children. Children's metacognitive judgments in their expertise areas are more mature than their judgments for unrelated skills in which they are not experts.

Traditionally, metacognitive development has been discussed in terms of domain-general cognitive advances in abilities such as executive processing, but it is not surprising that immersive experience in a particular skill or topic should also support more localized metacognitive growth. If you play thousands of games of chess, you will slowly build up a map of the success rates of openings or endgame configurations. While this may occur automatically and implicitly through statistical learning, it can lead to informed metacognitive judgments about the cognitive challenges posed by the task at hand. Similar metacognitive advances occur with successive explanatory refinements of a topic, such as how animals move. As explanatory prowess develops in an area, we gain a clearer sense of what we know, what gaps remain, and the nature of those gaps. Through a diverse range of such pockets of metacognitive precocity, children may also start to build more general intuitions of what it means for knowledge to be easy, difficult, or perhaps even impossible to attain. Offering children opportunities to gain distinctive areas of rich expertise provides cascades of cognitive benefits. They display a refreshing combination of humility and a willingness to question convention and dogma. Unless we are in situations where we must make immediate decisions, we should make such practices the norm.

6 Learn Where Ideas Come From

People should have a sense of how explanations of phenomena have unfolded over time. The history of ideas enables them to see that insight is an ongoing process of ever deeper causal understanding and that they are part of a long tradition of discoverers. Knowing a history of thought in an area can be both inspirational and profoundly informative to children well before they start formal schooling.

Elsewhere I've argued for the value of teaching current science through the historical origins of what we now know.[8] I won't repeat those arguments here beyond stating that a historical lens can greatly magnify the power of insight. To know why people struggled for centuries to harness the power of steam is to gain a much greater appreciation of what we know today. This benefit works for young children as well and provides an effective technique for making explanations more accessible to younger minds. By going back sufficiently far in history, the first steps of understanding may be easier to describe and envision. When the botanist Robert Brown first reported in 1827 how tiny pollen grains suspended in water jumped about irregularly, his observations led to key advances in thermodynamics fifty years later. Even later they led to Einstein's 1905 demonstration of how Brownian motion provided definitive evidence for the existence of atoms and molecules. A direct understanding of those later advances challenges most adults, but it is easy to grasp Brown's initial discovery and use it to see how thermodynamics and atomic theory evolved.

Historical accounts are also narrative in nature and, as such, have a different kind of appeal that can maintain engagement. When attempting to learn more about a new topic, I often find it especially effective to combine reading descriptions of the current state of the art with biographical accounts of early investigators who made key advances. In my attempts to understand modern AI systems I find myself turning more and more to descriptions of early pioneers in this area and how approaches have changed. These descriptions also help to remind us how much we tend to underestimate the extent to which we build upon the work of our predecessors.

7 Keep Humans in Meaningful Loops

Humans need to be in the loop of AI operations in ways that go beyond current practices. Many studies on loop inclusion with AIs ask humans to help other humans manage their feelings toward AIs, to judge the morality of AI activities, or simply to rate which AI features are most enjoyable or satisfying. These are useful ways to involve people, but the most important way to integrate humans is to have humans be part of the problem-solving process in order to improve AI performance and to make them as interpretable as possible. Humans matter because of our distinctive ways of dealing with novel cases, edge situations, idealizations, model-based reasoning, single case examples, and qualitative reasoning.

Humans generate revealing counterfactual scenarios not likely to emerge from AIs. For example, think about Maxwell's demon, the hypothetical nano-sized agent that opens and closes a gate depending on the speed of gas molecules. Current AI systems seem unable to dream up such an effective way of explaining randomness and entropy. We need to appreciate human cognition for its unique assets and not dismiss it as increasingly inferior to AI. We also need to keep different groups of humans involved in addressing XAI challenges. A diverse sampling of humans provides a wider range of informative counterfactuals and illustrates the value of simplifying a process in different ways. Novices and experts from a full range of disciplines may illustrate unique heuristics and strategies for handling complexity that were developed over extended periods of cultural and biological evolution. As AIs start to confront their own cases of computational overloads, they will benefit from adopting some of those techniques that have enabled humans to make major advances with limited cognitive resources. At the least, keeping humans in the loop will make it easier for other humans to understand what is happening inside those alien forms of intelligence.

* * *

We have repeatedly seen how insights allow us to see more deeply and act more effectively. These are intrinsically rewarding experiences that enrich our encounters with the world and give us a greater sense of efficacy. These alone are ample reasons for wanting to preserve opportunities for discovering the causal underpinnings of all that exists and happens around us.

But perhaps most important are the ways that understanding and explanation support two basic human rights: the right to know why and the right to know how. When we simply accept the supposedly inevitable rise of the unexplainable, we may be giving up far more than we first imagine.

The right to know why is widely endorsed but difficult to consistently respect. Every person is entitled to understand why decisions affecting them are made.[9] Setting aside limits imposed by factors such as community security and the ability to grasp even the simplest explanations (e.g., young infants), we see individuals and groups as obligated to explain to us the reasons behind actions that affect us. Whether it is being fired from a job, excluded from a press conference, or being urged to wear masks, decisions should be justified in an honest and transparent manner that allows, and even invites, further questions and discussion. Young children sense this right. When they request explanations for actions, we should support and nourish the exercising of that right in a reasonable manner. When machines undertake actions or make recommendations supposedly on our behalf, we should not easily give up the right to know why a machine came to that conclusion. We may choose to waive this right in special cases, but we should at least be aware of the possible consequences of doing so.

The right to know how is becoming more visible in areas such as the "right to repair" movement. We are entitled to have access to accurate and understandable explanations of how things operate. This includes not only the right to know how systems, technologies, and institutions work, but also the right to know when no one understands how something works. While acknowledging intellectual property and security concerns, we should not accept excessive use of those concerns to manipulate individuals. When cars show a sharp jump in prices because of the purported increase in expense of new technologies, we should be able to see whether those new costs were justified or were merely a pretext for increasing profits. If a computer company introduces a new model that no longer allows any user repairs and upgrades, we are entitled to object to this abridging of our agency unless it can be shown that the new model could not be economically produced without rendering previously serviceable components inaccessible. Enabling consumers to intervene on systems should be a manufacturer's default obligation, not a possible option.

In the end, we must appreciate how much deep insights are uniquely human gifts. Our dog, no matter how clever and manipulative she is, has no rights to know why and how—but our grandchildren do, even those still in infancy. Only humans experience the thrill of insight and the pleasures of seeing so much more, at least for now.

Acknowledgments

As I have for every other book I've written, I start with my deepest thanks to Kristi, my ever-forgiving partner for more than fifty-three magical years. But with this book, my gratitude is the greatest and most all-encompassing. As a brilliant scholar, gifted writer, and award-winning teacher, Kristi has patiently read through several drafts of the full manuscript, gently pointing out convoluted sections, missing examples, and awkward transitions. She has also been my most important and longest-lasting research collaborator. Her intuitions about clean experimental designs and about children's minds are unparalleled and have made all the difference on countless occasions. I am constantly amazed and humbled at her skills as a parent and grandparent; I have learned more from watching her interact with children than I have from decades of research.

My gratitude to Kristi goes even deeper just because this book has become so all-consuming in ways that neither of us had anticipated. Shortly after my last book on wonder came out, I could not help but think about its implications for how we gain insights into the causal underpinnings of reality. Before long, my total immersion in this project took over my life. Even though this preoccupation has invariably led to fewer shared experiences in recent years, Kristi has cheerfully kept supporting and encouraging me throughout. I am eagerly looking forward to a steady stream of shared adventures in the years ahead.

I am grateful in so many ways to our sons Derek, Dylan, and Marty. I thank them for being able to be a part of their exhilarating life journeys. They all have contributed to this book through many discussions and through penetrating comments on book sections. Their partners Eden, Kate, and Marita have also provided many thoughtful observations. Together, they have brought into this world our ever-inquisitive granddaughters Frances,

Maren, Poppy, and Willa, who constantly surprise and delight me with new insights about children and development.

Outside my family, I am grateful to an amazing array of people who have helped shape this project. I will be forever in debt to my friend and inspirational colleague, Paul Bloom. For years, Paul has constantly encouraged me to keep working on this book and has discussed every part of it in weekly zoom chats since its inception. His many comments on an earlier draft of this book were invaluable. I'm afraid I've been far less helpful in return. I have also learned much from several discussions with George Newman, who was working on his own book and was continuously generous with ideas and suggestions.

I will always be indebted to all members of my extended lab group, from postdocs to undergraduates to lab managers to summer interns. There are too many to list here, but they have all been important colleagues who have greatly shaped my ideas and research projects over the years. Whenever I describe a study using the pronoun *we*, I am indicating the involvement of members of that group, often in key roles. Their names are well represented in the many endnotes. I also thank several dozens of colleagues here at Yale and elsewhere who have chatted with me about many aspects of this project over the years.

I thank anonymous reviewers of an early stage of this book, whose suggestions greatly helped guide my writing of the first draft. I am especially indebted to two reviewers of a full draft of the book manuscript who made immensely useful and detailed comments; and I am delighted to be able to explicitly thank the computer scientist Mark Keane whose identity was revealed to me in the course of the review process.

Some of the material in this book is based on work supported by the National Science Foundation under (NSF grant DRL 1561143). Any opinions, findings, and conclusions or recommendations expressed in this material are those of the author and do not necessarily reflect the views of the National Science Foundation. I am also grateful to the National Institutes of Health, the Templeton Foundation, and the McDonnell Foundation for funding my research and interdisciplinary meetings. As with NSF, my writings here do not necessarily reflect the views of those funders.

Finally, I thank many people at the MIT Press. Ever since my undergraduate days at MIT, the press has been an essential part of my career growth as a cognitive scientist. Press director Amy Brand and senior editor

Phil Laughlin have been longtime colleagues in other projects, and I was delighted when they decided to support this book. Phil has been terrific in his choice of the reviewers and in his feedback on various drafts. I am grateful to several other people at MIT Press. I thank Haley Biermann for help with illustrations and book layout and Ginny Crossman and Laura Keeler for reading the manuscript with such care and wisdom. I also thank Katie Kerr for art preparation, Graham Atherton for cover design, Kate Elwell for production, and Helen Weldon for publicity. Finally, I thank my skilled indexer Tobiah Waldron.

Notes

Chapter 1

1. Campbell, Lewis (1882). *The Life of James Clerk Maxwell: With a Selection from His Correspondence and Occasional Writings and a Sketch of His Contributions to Science* (1st ed.). London: Macmillan; see also Lindsay, R. B. (1932). The broad point of view in physics. *Scientific Monthly, 34*(2), 115–124.

2. Dreyfus, C. (2019, January 24). "The joy of the discovery": An interview with Jennifer Doudna. *New York Review of Books.* https://www.nybooks.com/online/2019/01/24/the-joy-of-the-discovery-an-interview-with-jennifer-doudna; Pillock, A. (2015, May 11). Jennifer Doudna, a pioneer who helped simplify genome editing. *New York Times.*

3. Keil, F. C. (2022). *Wonder: Childhood and the Lifelong Love of Science.* MIT Press.

4. Kohler, W. (2018/1917). *The Mentality of Apes.* Routledge. Others have focused on the meaning of insight where suddenness is central; e.g., Topolinski, S., & Reber, R. (2010). Gaining insight into the "Aha" experience. *Current Directions in Psychological Science, 19*(6), 402–405.

5. E.g., Hatcher, R. L. (1973). Insight and self-observation. *Journal of the American Psychoanalytical Association, 21*, 377–398.

6. Forbes, N., & Mahon, B. (2014). *Faraday, Maxwell, and the Electromagnetic Field: How Two Men Revolutionized Physics.* New York: Prometheus Books.

7. Tweney, R. D. (1991). Faraday's notebooks: The active organization of creative science. *Physics Education, 26*(5), 301; Tweney, R. D. (1989). Fields of enterprise: On Michael Faraday's thought. In Wallace, D. B., & Gruber, H. E. (Eds.), *Creative People at Work: Twelve Cognitive Case Studies* (pp. 91–106). Oxford University Press.

8. Darwin Correspondence Project. (n.d.). Darwin's species notebooks: "I think . . ." University of Cambridge. Retrieved August 1, 2025, from https://www.darwinproject.ac.uk/commentary/evolution/darwin-s-species-notebooks-i-think

9. See also Basalla, G. (1989). *The Evolution of Technology*. Cambridge University Press.

10. E.g., Burback, L., Brult-Phillips, S., Nijdam, M. J., McFarlane, A., & Vermetten, E. (2024). Treatment of posttraumatic stress disorder: a state-of-the-art review. *Current Neuropharmacology, 22*(4), 557–635.

11. Naveau A., Seidel K., & Klein O. D. (2014). Tooth, hair and claw: Comparing epithelial stem cell niches of ectodermal appendages. *Experimental Cell Research, 325*(2), 96–103.

12. Kaiho, K., & Oshima, N. (2017). Site of asteroid impact changed the history of life on Earth: The low probability of mass extinction. *Scientific Reports 7*(1), 14855.

13. E.g., Tanenhaus, M. K., Carlson, G., & Trueswell, J. C. (1989). The role of thematic structures in interpretation and parsing. *Language and Cognitive Processes, 4*(3–4), SI211–SI234.

14. Loewenstein, G. (1994). The psychology of curiosity: A review and reinterpretation. *Psychological Bulletin, 116*(1), 75–98. https//doi.org/10.1037/0033-2909.116.1.75; Liquin, E. G., & Lombrozo, T. (2020). Explanation-seeking curiosity in childhood. *Current Opinion in Behavioral Sciences, 35*, 14–20; Murayama, K. (2022). A reward-learning framework of knowledge acquisition: An integrated account of curiosity, interest, and intrinsic–extrinsic rewards. *Psychological Review, 129*(1), 175; Poli, F., O'Reilly, J. X., Mars, R. B., & Hunnius, S. (2024). Curiosity and the dynamics of optimal exploration. *Trends in Cognitive Sciences, 28*(5), 441–453.

15. Thesen, A., Steen, J. B., & Døving, K. B. (1993). Behaviour of dogs during olfactory tracking. *Journal of Experimental Biology, 180*(1), 247–251.

16. Keil. *Wonder*.

17. Plato. (1972). *Phaedrus* (R. Hackforth, Trans.). Cambridge University Press.

18. For a more extended discussion of such earlier warnings, see Thompson, C. (2014). *Smarter Than You Think: How Technology Is Changing Our Minds for the Better*. Penguin.

19. Smiley, A. H., & Fisher, M. (2022). The golden age is behind us: How the status quo impacts the evaluation of technology. *Psychological Science, 33*(9), 1605–1614.

20. Cummings, M. L., Gao, F., & Thornburg, K. M. (2016). Boredom in the workplace: A new look at an old problem. *Human Factors, 58*(2), 279–300; Mueller, G. (2021). *Breaking Things at Work: The Luddites Are Right About Why You Hate Your Job*. Verso Books.

21. Lundstrom, M. S., & Alam, M. A. (2022). Moore's law: The journey ahead. *Science, 378*(6621), 722–723.

Chapter 2

1. Clark-Hachtel, C. M., Hibshman, J. D., De Buysscher, T., Stair, E. R., Hicks, L. M., & Goldstein, B. (2024). The tardigrade *Hypsibius exemplaris* dramatically upregulates DNA repair pathway genes in response to ionizing radiation. *Current Biology, 34*(9), 1819–1830.

2. de Solla Price, D. (1974). Gears from the Greeks: The Antikythera mechanism, a calendar computer from ca. 80 BC. *Transactions of the American Philosophical Society*, 1–70; Seiradakis, J. H., & Edmunds, M. G. (2018). Our current knowledge of the Antikythera Mechanism. *Nature Astronomy, 2*(1), 35–42.

3. Chi, M. T., De Leeuw, N., Chiu, M. H., & LaVancher, C. (1994). Eliciting self-explanations improves understanding. *Cognitive Science, 18*(3), 439–477.

4. Zee, A. (1999). *Fearful Symmetry: The Search for Beauty in Modern Physics*. Princeton University Press.

5. Vosniadou, S., & Brewer, W. F. (1992). Mental models of the Earth: A study of conceptual change in childhood. *Cognitive Psychology, 24*(4), 535–585.

6. Lombrozo, T. (2007). Simplicity and probability in causal explanation. *Cognitive Psychology, 55*(3), 232–257; Johnson, S. G., Valenti, J. J., & Keil, F. C. (2019). Simplicity and complexity preferences in causal explanation: An opponent heuristic account. *Cognitive Psychology, 113*, 101222.

7. Zemla, J., Sloman, S. A., Bechlivanidis, C., & Lagnado, D. (2023). Not so simple! Mechanisms increase preference for complex explanations. *Cognition, 239*. https://doi.org/10.1016/j.cognition.2023.105551

8. Lombrozo, T. (2016). Explanatory preferences shape learning and inference. *Trends in Cognitive Sciences, 20*(10), 748–759.

9. McCarthy, A. M., & Keil, F. C. (2023). A right way to explain? Function, mechanism, and the order of explanations. *Cognition, 238*, 105494; McCarthy, A. M., Betz, N., & Keil, F. C. (2024). Adults' learning of complex explanations violates their intuitions about optimal explanatory order. *Cognition, 246*, 105767.

10. McCarthy, A. M. (2024). Exploring function and mechanism through intuitions, explanations, & learning. Ph.D. dissertation, Yale University.

11. Bechtel, W. (2011). Mechanism and biological explanation. *Philosophy of Science, 78*(4), 533–557; Bechtel, W., & Bollhagen, A. (2024). Causal cycles in biology. In Illari, P., & Russo, F. (Eds.), *The Routledge Handbook of Causality and Causal Methods* (pp. 255–268). Routledge.

12. Gelman, S. A. (2003). *The Essential Child: Origins of Essentialism in Everyday Thought*. Oxford University Press; Keil, F. C. (1989). *Concepts, Kinds and Cognitive Development*. MIT Press.

13. Gelman, S. A. (2023). Looking beyond the obvious. *American Psychologist, 78*(5), 667.

14. Gelman, S. A., & Wellman, H. M. (1991). Insides and essences: Early understandings of the non-obvious. *Cognition, 38*(3), 213–244.

15. Leslie, S. J. (2013). Essence and natural kinds: When science meets preschooler intuition. *Oxford Studies in Epistemology, 4*, 108–165.

16. Weisberg, D. S., Keil, F. C., Goodstein, J., Rawson, E., & Gray, J. R. (2008). The seductive allure of neuroscience explanations. *Journal of Cognitive Neuroscience, 20*(3), 470–477.

17. Rozenblit, L., & Keil, F. (2002). The misunderstood limits of folk science: An illusion of explanatory depth. *Cognitive Science, 26*(5), 521–562.

18. For examples see Alter, A. L., Oppenheimer, D. M., & Zemla, J. C. (2010). Missing the trees for the forest: A construal level account of the illusion of explanatory depth. *Journal of Personality and Social Psychology, 99*(3), 436; Vitriol, J. A., & Marsh, J. K. (2018). The illusion of explanatory depth and endorsement of conspiracy beliefs. *European Journal of Social Psychology, 48*(7), 955–969; Mills, C. M., & Keil, F. C. (2004). Knowing the limits of one's understanding: The development of an awareness of an illusion of explanatory depth. *Journal of Experimental Child Psychology, 87*(1), 1–32; Meyers, E. A., Gretton, J. D., Budge, J. R., Fugelsang, J. A., & Koehler, D. J. (2023). Broad effects of shallow understanding: Explaining an unrelated phenomenon exposes the illusion of explanatory depth. *Judgment and Decision Making, 18*, e24; Fernbach, P. M., Sloman, S. A., Louis, R. S., & Shube, J. N. (2013). Explanation fiends and foes: How mechanistic detail determines understanding and preference. *Journal of Consumer Research, 39*(5), 1115–1131; Ylikoski, P. K. (2009). The illusion of depth of understanding in science. In de Regt, H. W., Leonelli, S., & Eigner, K. (Eds.), *Scientific Understanding: Philosophical Perspectives* (pp. 100–119). University of Pittsburgh Press.

19. Lawson, R. (2006). The science of cycology: Failures to understand how everyday objects work. *Memory & Cognition, 34*(8), 1667–1675.

20. These are also patent categories in in the International Patent Classification system, or IPC.

21. Chromik, M., Eiband, M., Buchner, F., Krüger, A., & Butz, A. (2021, April). I think I get your point, AI! The illusion of explanatory depth in explainable AI. In *Proceedings of the 26th International Conference on Intelligent User Interfaces* (pp. 307–317). Association for Computing Machinery. https://doi.org/10.1145/3397481.3450644

22. Weisberg, M. (2007). Three kinds of idealization. *Journal of Philosophy, 104*(12), 639–659; Levy, A. (2021). Idealization and abstraction: Refining the distinction. *Synthese, 198*, 5855–5872.

23. E.g., Liu, R., Dou, Z., Tian, T., Gao, X., Chen, L., Yuan, X., . . . & Liu, X. (2023). Dynamic phosphorylation of CENP-N by CDK1 guides accurate chromosome segregation in mitosis. *Journal of Molecular Cell Biology*, mjad041.

24. Sikora, M., Bohdal, T., & Kruzel, M. (2023). High pressure refrigerants condensation modeling in horizontal pipe minichannels. *Heat Transfer Engineering, 44*(11–12), 970–986.

25. Harris, P. L., & Koenig, M. A. (2006). Trust in testimony: How children learn about science and religion. *Child Development, 77*(3), 505–524; Harris, P. L., Koenig, M. A., Corriveau, K. H., & Jaswal, V. K. (2018). Cognitive foundations of learning from testimony. *Annual Review of Psychology, 69*(1), 251–273; Gweon, H. (2021). Inferential social learning: Cognitive foundations of human social learning and teaching. *Trends in Cognitive Sciences, 25*(10), 896–910.

26. Rabb, N., Fernbach, P. M., & Sloman, S. A. (2019). Individual representation in a community of knowledge. *Trends in Cognitive Sciences, 23*(10), 891–902.

27. Wegner, D. M. (1987). Transactive memory: A contemporary analysis of the group mind. In Mullen, B., & Goethals, G. R. (Eds.), *Theories of Group Behavior* (pp. 185–208). Springer.

28. Hutchins, E. (1995). *Cognition in the Wild*. MIT Press.

29. Richardson, E., Sheskin, M., & Keil, F. C. (2021). An illusion of self-sufficiency for learning about artifacts in scaffolded learners, but not observers. *Child Development, 92*(4), 1523–1538.

30. Wilson, R. A., & Keil, F. (1998). The shadows and shallows of explanation. *Minds and Machines, 8*, 137–159.

31. Radvansky, G. A., & Copeland, D. E. (2006). Walking through doorways causes forgetting: Situation models and experienced space. *Memory & Cognition, 34*, 1150–1156; Zacks, J. M. (2020). Event perception and memory. *Annual Review of Psychology, 71*(1), 165–191; Radvansky, G. A., & Zacks, J. M. (2017). Event boundaries in memory and cognition. *Current Opinion in Behavioral Sciences, 17*, 133–140.

32. Hume D. (1748). *Philosophical Essays Concerning Human Understanding*. Georg Olms Verlag.

33. Byrne, R. M. (2016). Counterfactual thought. *Annual Review of Psychology, 67*(1), 135–157; Gerstenberg, T. (2024). Counterfactual simulation in causal cognition. *Trends in Cognitive Sciences, 28*, 924–936.

34. Gamow, G. (1940). *Mr. Tompkins in Wonderland* (pp. 2–3). Cambridge University Press.

35. Keil, F. C., Smith, W. C., Simons, D. J., & Levin, D. T. (1998). Two dogmas of conceptual empiricism: Implications for hybrid models of the structure of knowledge. *Cognition, 65*(2–3), 103–135.

36. Lewis, D. (2013). *Counterfactuals*. John Wiley & Sons.

37. Byrne, R. M. (2016). Counterfactual thought. *Annual Review of Psychology, 67*(1), 135–157.

38. Craver, C., Tabery, J., & Illari, P. (2024). Mechanisms in science. In Zalta, E. N., & Nodelman, U. (Eds.), *The Stanford Encyclopedia of Philosophy*. https://plato.stanford.edu/archives/fall2024/entries/science-mechanisms/

39. Keil, F. C., & Lockhart, K. L. (2021). Beyond cause: The development of clockwork cognition. *Current Directions in Psychological Science, 30*(2), 167–173; Chuey, A., Lockhart, K., Trouche, E., & Keil, F. (2023). Young children can infer information preferences from goals and recommend appropriate sources to others. *Developmental Psychology, 59*(8), 1511; Chuey, A., McCarthy, A., Lockhart, K., Trouche, E., Sheskin, M., & Keil, F. (2021). No guts, no glory: Underestimating the benefits of providing children with mechanistic details. *npj Science of Learning, 6*(1), 30.

40. Craver, Tabery, & Illari. Mechanisms in science; Machamer, P., Darden, L., & Craver, C. F. (2000). Thinking about mechanisms. *Philosophy of Science, 67*(1), 1–25; Bechtel, W., & Abrahamsen, A. (2005). Explanation: A mechanist alternative. *Studies in History and Philosophy of Science Part C: Studies in History and Philosophy of Biological and Biomedical Sciences, 36*(2), 421–441.

41. Bechtel, W. (2011). Mechanism and biological explanation. *Philosophy of Science, 78*(4), 533–557.

42. Mack, K. (2021). *The End of Everything (Astrophysically Speaking)*. Simon and Schuster.

43. Franklin, A., & Fischbach, E. (2016). The rise and fall of the fifth force: Discovery, pursuit, and justification in modern physics. Springer; NASA Space News. (2025, January 12). Could a New Force Rewrite Physics? The Fifth Force and the Dark Universe. https://nasaspacenews.com/2025/01/could-a-new-force-rewrite-physics-the-fifth-force-and-the-dark-universe/

44. Boyd, R. (1999). Homeostasis, species, and higher taxa. *Species: New Interdisciplinary Essays, 141*, 185; Boyd, R. (2010). Homeostasis, higher taxa, and monophyly. *Philosophy of Science, 77*(5), 686–701; Boyd, R. (2021). Rethinking natural kinds, reference and truth: Towards more correspondence with reality, not less. *Synthese, 198*(Suppl. 12), 2863–2903.

45. Mou, W. (2025). Representing place locations and orientations in cognitive maps. *Nature Reviews Psychology, 4*, 347–360; Freas, C. A., & Cheng, K. (2022). The basis of navigation across species. *Annual Review of Psychology, 73*(1), 217–241; Jeffery, K. J., Cheng, K., Newcombe, N. S., Bingman, V. P., & Menzel, R. (2024). Unpacking the navigation toolbox: Insights from comparative cognition. *Proceedings of the Royal Society B, 291*(2016), 20231304.

46. Huey, H., Lu, X., Walker, C. M., & Fan, J. E. (2023). Visual explanations prioritize functional properties at the expense of visual fidelity. *Cognition, 236*, 105414.

47. Heiser, J., & Tversky, B. (2006). Arrows in comprehending and producing mechanical diagrams. *Cognitive Science, 30*(3), 581–592.

48. Boger, T., & Keil, F. C. (2025). Mechanistic complexity is fundamental: Evidence from judgments, attention, and memory. *Journal of Experimental Psychology: General, 154*(2), 596–606. https://doi.org/10.1037/xge0001691

49. Propp, V. (1968). *Morphology of the Folktale*. University of Texas Press.

50. Mandler, J. M. (1982). Recent research on story grammars. *Advances in Psychology, 9*, 207–218.

51. Sterelny also suggests this idea in his book *The Evolved Apprentice* (MIT Press, 2012) with an example of color-coded wires.

52. Simon, H. (1969). *Sciences of the Artificial*. MIT Press.

53. Alexander, R. M. (2005). Mechanics of animal movement. *Current Biology, 15*(16), R616–R619; Vogel, S. (1998). *Cat's Paws and Catapults: Mechanical Worlds of Nature and People*. W. W. Norton.

54. Kelemen, D., & Rosset, E. (2009). The human function compunction: Teleological explanation in adults. *Cognition, 111*(1), 138–143.

55. Craver, C. F. (2013). Functions and mechanisms: A perspectivalist view. In *Functions: Selection and Mechanisms* (pp. 133–158). Springer.

56. Mayr, E. (1961). Cause and effect in biology: Kinds of causes, predictability, and teleology are viewed by a practicing biologist. *Science, 134*(3489), 1501–1506.

57. Gould, S. J. (2010). *The Panda's Thumb: More Reflections in Natural History*. W. W. Norton.

58. Artifacts don't always backtrack and update, as shown by the case of the qwerty keyboard (thanks to Paul Bloom for this example), but they backtrack and update much more than organisms.

59. Trowsdale, J. (2024). *What the Body Knows: A Guide to the New Science of Our Immune System*. Yale University Press; see also Gerhart, J., & Kirschner, M. (2007). The theory of facilitated variation. *Proceedings of the National Academy of Sciences, 104*(suppl. 1), 8582–8589.

60. Bechtel, W., & Bich, L. (2021). Grounding cognition: Heterarchical control mechanisms in biology. *Philosophical Transactions of the Royal Society B, 376*(1820), 20190751.

61. E.g., Paulson, J. R., Hudson, D. F., Cisneros-Soberanis, F., & Earnshaw, W. C. (2021). Mitotic chromosomes. *Seminars in Cell & Developmental Biology, 117*, 7–29; Nielsen, C. F., Zhang, T., Barisic, M., Kalitsis, P., & Hudson, D. F. (2020). Topoisomerase IIα is essential for maintenance of mitotic chromosome structure. *Proceedings of the National Academy of Sciences, 117*(22), 12131–12142.

62. Pollard, T. D., & Cooper, J. A. (2009). Actin, a central player in cell shape and movement. *Science, 326*(5957), 1208–1212; Lappalainen, P., Kotila, T., Jégou, A.,

& Romet-Lemonne, G. (2022). Biochemical and mechanical regulation of actin dynamics. *Nature Reviews Molecular Cell Biology, 23*(12), 836–852.

63. Yao, N., Zhang, Z., Yu, L., Hazarika, R., Yu, C., Jang, H., . . . & Johannes, F. (2023). An evolutionary epigenetic clock in plants. *Science, 381*(6665), 1440–1445.

64. Kearnes, M., Kuch, D., & Johnston, A. (2018). How to do things with metaphors: Engineering life as hodgepodge. *Life Sciences, Society and Policy, 14*, 1–17.

65. Boudry, M., & Pigliucci, M. (2013). The mismeasure of machine: Synthetic biology and the trouble with engineering metaphors. *Studies in History and Philosophy of Science Part C: Studies in History and Philosophy of Biological and Biomedical Sciences, 44*(4), 660–668.

66. Ross, L. N., & Bassett, D. S. (2024). Causation in neuroscience: Keeping mechanism meaningful. *Nature Reviews Neuroscience, 25*, 81–90.

67. Nicholson, D. J. (2019). Is the cell really a machine? *Journal of Theoretical Biology, 477*, 108–126.

68. Levy, A. (2023). Molecular-biological machines: A defense. *Biology & Philosophy, 38*(5), 1–19; Santiveri, M., Roa-Eguiara, A., Kühne, C., Wadhwa, N., Hu, H., Berg, H. C., . . . & Taylor, N. M. (2020). Structure and function of stator units of the bacterial flagellar motor. *Cell, 183*(1), 244–257.

Chapter 3

1. Ryan, J. (2018). *Race to Hawaii: The 1927 Dole Air Derby and the Thrilling First Flights that Opened the Pacific*. Chicago Review Press.

2. Gladwin, T., & Gladwin, T. (2009). *East Is a Big Bird: Navigation and Logic on Puluwat Atoll*. Harvard University Press; Thomas, N. (2021). *Voyagers: The Settlement of the Pacific*. Bloomsbury.

3. Bhattarai, T., Ebong, A., & Raja, M. Y. A. (2024). A review of light-emitting diodes and ultraviolet light-emitting diodes and their applications. *Photonics, 11*(6), 491.

4. See https://saving-old-seagulls.co.uk/index.html.

5. Stamp, J. (2013, January 3). The QWERTY keyboard will never die. Where did the 150-year-old design come from? *Smithsonian Magazine*. https://www.smithsonianmag.com/arts-culture/fact-of-fiction-the-legend-of-the-qwerty-keyboard-49863249/

6. Danezis, G., & Clayton, R. (2007). Introducing traffic analysis. In Acquisti, A., Gritzalis, S., Lambrinoudakis, C., & di Vimercati, S. (Eds.), *Digital Privacy: Theory, Technologies, and Practices* (pp. 95–116). Auerbach Publications.

7. Bohn, Dieter. (2020, May 4). The saga of Apple's bad butterfly MacBook keyboards is finally over. *The Verge*. https://www.theverge.com/2020/5/4/21246223

/macbook-keyboard-butterfly-magic-pro-apple-design; Ashworth, Boone. (2022, July 19). Apple's butterfly keyboard fiasco leads to a $50M settlement. *Wired.* https://www.wired.com/story/apple-butterfly-keyboard-settlement-50-million/

8. Ma, Z., Edge, D., Findlater, L., & Tan, H. Z. (2015). Haptic keyclick feedback improves typing speed and reduces typing errors on a flat keyboard. In *2015 IEEE World Haptics Conference* (pp. 220–227). IEEE.

9. Tommasini, A. (2020, November 12). Why do pianists know so little about pianos? *New York Times.*

10. Saitis, C., Järveläinen, H., & Fritz, C. (2018). *The Role of Haptic Cues in Musical Instrument Quality Perception* (pp. 73–93). Springer International.

11. Greve, L., & Dyson, S. (2020). What can we learn from visual and objective assessment of non-lame and lame horses in straight lines, on the lunge and ridden? *Equine Veterinary Education, 32*(9), 479–491; Dyson, S., Routh, J., Bondi, A., & Pollard, D. (2022). Gait abnormalities and ridden horse behaviour in a convenience sample of the United Kingdom ridden sports horse and leisure horse population. *Equine Veterinary Education, 34*(2), 84–95.

12. Shapiro, L. (2019). *Embodied Cognition.* Routledge; Gallagher, S. (2023). *Embodied and Enactive Approaches to Cognition.* Cambridge University Press.

13. Lescieux, C., Singh, A., & Doyon-Poulin, P. (2025). When it goes right, when it goes wrong: Lessons learned from automation in aviation. In Coursaris, C. K., Beringer, J., Léger, P. M., & Öz, B. (Eds.), *The Design of Human-Centered Artificial Intelligence for the Workplace* (pp. 329–350). Springer. https://doi.org/10.1007/978-3-031-83512-4_19; Nicolin, I., & Nicolin, B. A. (2019). The fly-by-wire system. *Incas Bulletin, 11*(4), 217–222.

14. Timmermans, S., Dehez, B., & Fisette, P. (2020). Multibody-based piano action: Validation of a haptic key. *Machines, 8*(4), 76.

15. Gibson, J. J. (2014). The theory of affordances. In Gieseking, J. J., Mangold, W., Katz, C., Low, S., & Saegert, S. (Eds.), *The People, Place, and Space Reader* (pp. 56–60). Routledge; Gibson, J. J. (2014). *The Ecological Approach to Visual Perception* (classic edition). Psychology Press.

16. Simon, H. A. (2019). *The Sciences of the Artificial,* reissue of the third edition with a new introduction by John Laird. MIT Press; Simon, H. (2000). Discovering explanations. In Keil, F. C., & Wilson, R. A. (Eds.), *Explanation and Cognition* (pp. 21–59). MIT Press.

17. Suleyman, M. (2023). *The Coming Wave: Technology, Power, and the Twenty-First Century's Greatest Dilemma.* Crown.

18. E.g., Packard, V. (1960). *The Waste Makers.* Penguin; Slade, G. (2009). *Made to Break: Technology and Obsolescence in America.* Harvard University Press.

19. Collett, J. P. (2013). The history of electronics: From vacuum tubes to transistors. In Krige, J., & Pestre, D. (Eds.), *Science in the Twentieth Century* (pp. 253–274). Routledge; Guarnieri, M. (2012). The age of vacuum tubes: Early devices and the rise of radio communications. *IEEE Industrial Electronics Magazine, 6*(1), 41–43.

20. Gentner, D., & Gentner, D. R. (2014). Flowing waters or teeming crowds: Mental models of electricity. In Gentner, D., & Stevens, A. L. (Eds.), *Mental Models* (pp. 99–129). Psychology Press.

21. Windsor, H. H. (Ed.). (1913). *The Boy Mechanic* (Vol. 1). Popular Mechanics.

22. Isaacson, Walter (2014, September 18). Walter Isaacson on the women of ENIAC. Fortune.

23. Tricker, R. A. R. (1966). *The Contributions of Faraday and Maxwell to Electrical Science*. Elsevier; Forbes, N., & Mahon, B. (2014). *Faraday, Maxwell, and the Electromagnetic Field: How Two Men Revolutionized Physics*. Prometheus Books.

24. Clegg, B. (2019). *Professor Maxwell's Duplicitous Demon: The Life and Science of James Clerk Maxwell*. Icon Books.

25. Multiple models were created in 1874, with one copy sent to Gibbs at Yale. For a description of the Yale model see https://peabodyhsi.wordpress.com/2020/07/15/3d-scanning-the-famous-maxwell-gibbs-thermodynamic-model/

26. Berreby, D. (2025, January 27). On AI's blank face, we project our psyche. *Robots for the rest of us* (blog). https://robots4therestofus.substack.com/p/on-ais-blank-face-we-project-our

Chapter 4

1. Salecker, G. E. (2022). *Destruction of the Steamboat Sultana: The Worst Maritime Disaster in American History*. Naval Institute Press; Nichols, D. (1996). ASME boiler code became constitution for steam age. *Power Engineering, 100*(2), 27–31; Varrasi, J. (2005). The true harnessing of steam. *Mechanical Engineering, 127*(1), 46–48; Burke, J. G. (1966). Bursting boilers and the federal power. *Technology and Culture, 7*(1), 1–23.

2. For more on the challenges of engineering steam power see Dickinson, H. W. (2022). *A Short History of the Steam Engine*. Routledge; Sen, P. (2021). *Einstein's Fridge: How the Difference Between Hot and Cold Explains the Universe*. Simon and Schuster; Hammack, B. (2023). *The Things We Make: The Unknown History of Invention from Cathedrals to Soda Cans*. Sourcebooks.

3. Henrich, J., & Muthukrishna, M. (2024). What makes us smart? *Topics in Cognitive Science, 16*(2), 322–342.

4. Muthukrishna, M., & Henrich, J. (2016). Innovation in the collective brain. *Philosophical Transactions of the Royal Society B: Biological Sciences, 371*(1690), 20150192;

Legare, C. H. (2017). Cumulative cultural learning: Development and diversity. *Proceedings of the National Academy of Sciences, 114*(30), 7877–7883.

5. Muthukrishna, M., Doebeli, M., Chudek, M., & Henrich, J. (2018). The cultural brain hypothesis: How culture drives brain expansion, sociality, and life history. *PLoS Computational Biology, 14*(11), e1006504.

6. Markov, A. V., & Markov, M. A. (2020). Runaway brain-culture coevolution as a reason for larger brains: Exploring the "cultural drive" hypothesis by computer modeling. *Ecology and Evolution, 10*(12), 6059–6077; Vandervert, L., Manto, M., Adamaszek, M., Ferrari, C., Ciricugno, A., & Cattaneo, Z. (2024). The evolution of the optimization of cognitive and social functions in the cerebellum and thereby the rise of *Homo sapiens* through cumulative culture. *The Cerebellum 23*(5), 1981–1992.

7. Boyd, R., & Richerson, P. J. (2005). *The Origin and Evolution of Cultures*. Oxford University Press.

8. Henrich & Muthukrishna. What makes us smart?, 329.

9. Caspari, R. (2011). The evolution of grandparents. *Scientific American 305*(2), 44–49; Danielsbacka, M., & Tanskanen, A. O. (2021). Grandparents and grandchildren. In Shackelford, T. K., & Weekes-Shackelford, V. A. (Eds.), *Encyclopedia of Evolutionary Psychological Science* (pp. 3523–3528). Springer International.

10. Lew-Levy, S., & Amir, D. (2024). Children as agents of cultural adaptation. *Behavioral and Brain Sciences*, 1–68.

11. Diamond, J. M. (1978). The Tasmanians: The longest isolation, the simplest technology. *Nature, 273*(5659), 185–186; Henrich, J. (2004). Demography and cultural evolution: How adaptive cultural processes can produce maladaptive losses—the Tasmanian case. *American Antiquity, 69*(2), 197–214.

12. Read, D. (2006). Tasmanian knowledge and skill: Maladaptive imitation or adequate technology? *American Antiquity, 71*(1), 164–184.

13. Bebber, M. R. (2021). The role of functional efficiency in the decline of North America's copper culture (8000–3000 BP): An experimental, ecological, and evolutionary approach. *Journal of Archaeological Method and Theory, 28*(4), 1224–1260.

14. Dowling, P. (2019). The Khmer Rouge destroyed education in Cambodia—now the country is fighting back. *The Independent*. https://www.independent.co.uk/arts-entertainment/photography/khmer-rouge-education-cambodia-ngo-pol-pot-a9233851.html; Chigas, G., & Mosyakov, D. (2001). Literacy and education under the Khmer Rouge. Genocide Studies Program, MacmIllan Center for International and Area Studies at Yale, Yale University. https://macmillan.yale.edu/gsp/literacy-and-education-under-khmer-rouge

15. Stanchevici, D. (2017). *Stalinist Genetics: The Constitutional Rhetoric of TD Lysenko*. Routledge; Kean, S. (2017, December 19). The Soviet era's deadliest scientist is regaining popularity in Russia. *The Atlantic*.

16. Keil, F. C., Stein, C., Webb, L., Billings, V. D., & Rozenblit, L. (2008). Discerning the division of cognitive labor: An emerging understanding of how knowledge is clustered in other minds. *Cognitive Science, 32*(2), 259–300; Fisher, M., & Oppenheimer, D. M. (2021). Who knows what? Knowledge misattribution in the division of cognitive labor. *Journal of Experimental Psychology: Applied, 27*(2), 292–306; Sloman, S., & Fernbach, P. (2018). *The Knowledge Illusion: Why We Never Think Alone*. Penguin.

17. Osiurak, F., Claidière, N., Bluet, A., Brogniart, J., Lasserre, S., Bonhoure, T., . . . & Reynaud, E. (2022). Technical reasoning bolsters cumulative technological culture through convergent transformations. *Science Advances, 8*(9), eabl7446; Osiurak, F., Bryche, C., Bluet, A., & Reynaud, E. (2024). From technical reasoning to cumulative technological culture. In Wynn, T., Overmann, K. A., & Coolidge, F. L. (Eds.), *The Oxford Handbook of Evolutionary Cognitive Archaeology* (pp. 173–196). Oxford University Press.

18. Davis, N. (2014, September 21). Interview: Maggie Aderin-Pocock: How a space-obsessed schoolgirl battled the odds to become a top scientist. *The Guardian*. https://www.theguardian.com/science/2014/sep/21/maggie-aderin-pocock-interview-bbc-nasa-space

19. Smith, N. (2019). Interview: Maggie Aderin-Pocock: Words and portrait. *Engineering & Technology, 14*(6), 74–77.

20. Rots, V. (2016). Projectiles and hafting technology. In Iovita, R., & Sano, K. (Eds.), *Multidisciplinary Approaches to the Study of Stone Age Weaponry* (pp. 167–185). Springer; Barham, L. (2013). *From Hand to Handle: The First Industrial Revolution*. Oxford University Press.

21. Barham. *From Hand to Handle*.

22. Markov, A. V., & Markov, M. A. (2020). Runaway brain-culture coevolution as a reason for larger brains: Exploring the "cultural drive" hypothesis by computer modeling. *Ecology and Evolution, 10*(12), 6059–6077.

23. Agrawal, R. (2023). *Nuts and Bolts: Seven Small Inventions That Changed the World in a Big Way*. W. W. Norton.

24. See, for example, Lunteren, F. V. (2016). Clocks to computers: A machine-based "big picture" of the history of modern science. *Isis, 107*(4), 762–776; Riskin, J. (2020). *The Restless Clock: A History of the Centuries-Long Argument Over What Makes Living Things Tick*. University of Chicago Press.

25. Roudaut, S. (2022). Clocks, automata and the mechanization of nature (1300–1600). *Philosophies, 7*(6), 139.

26. Dolnick, E. (2011). *The Clockwork Universe*. HarperCollins; Lunteren. Clocks to computers.

27. Laudan, L. (1966). The clock metaphor and probabilism: The impact of Descartes on English methodological thought, 1650–1665. *Annals of Science*, *22*(2), 73–104.

28. Dolnick. *The Clockwork Universe*.

29. Wood, G. (2002, February 15). Living dolls: A magical history of the quest for mechanical life by Gaby Wood. *The Guardian*. https://www.theguardian.com/books/2002/feb/16/extract.gabywood; Byrd, M. (2012). Man as Machine. *Wilson Quarterly (1976–)*, *36*(1), 30–34.

30. Riskin, J. (2003). The defecating duck, or, the ambiguous origins of artificial life. *Critical Inquiry*, *29*(4), 599–633; Voltaire. 'Discours en vers sur l'homme' (1738). In *Oeuvres completes*, 10 vols. (Paris, 1877), 9:420; Voskuhl, A. (2019). *Androids in the Enlightenment: Mechanics, Artisans, and Cultures of the Self.* University of Chicago Press.

31. Pynchon, T. (2012). *Mason & Dixon*. Penguin.

32. Baida, P. (1987). Eli Whitney's other talent. *American Heritage*, *38*(4), 22–23; Woodbury, R. S. (1960). The legend of Eli Whitney and interchangeable parts. *Technology and Culture*, *1*(3), 235–253.

33. Wimsatt, W. C. (2013). Entrenchment and scaffolding: Developing scaffolds in evolution, culture, and cognition. In Caporael, L. R., Griesemer, J. R., & Wimsatt, W. C. (Eds.), *Developing Scaffolds in Evolution, Culture, and Cognition* (pp. 77–106). MIT Press.

34. Herkimer, H. (1952). *Engineers' Illustrated Thesaurus*. Chemical Publishing Company; Kendall, J. (2008). *The Man Who Made Lists: Love, Death, Madness, and the Creation of Roget's Thesaurus*. Penguin.

35. Singh, A. (2017). *Fundamentals of Machine Design* (Vol. 2). Cambridge University Press; Koetsier, T. (2023). A note on the history of the science of machines. In Ceccarelli, E., & Jauregui-Correa, J. C. (Eds.), *State-of-the-Art and Innovations in Mechanism and Machine Science: A Tribute to Carlos López-Cajún* (pp. 31–45). Springer Nature.

36. Riede, F., Lew-Levy, S., Johannsen, N. N., Lavi, N., & Andersen, M. M. (2023). Toys as teachers: A cross-cultural analysis of object use and enskillment in hunter–gatherer societies. *Journal of Archaeological Method and Theory*, 30(1), 32–63.

37. Healey, A., Mendelsohn, A., Sells, J. M., Donoghue, E., Earls, M., Hashikawa, A., . . . & Williams, P. G. (2019). Selecting appropriate toys for young children in the digital era. *Pediatrics*, *143*(1), e20183348.

38. E.g., Venker, C. E., & Johnson, J. R. (2022). Electronic toys decrease the quantity and lexical diversity of spoken language produced by children with autism spectrum

disorder and age-matched children with typical development. *Frontiers in Psychology, 13*, 929589.

39. Boudry, M., & Pigliucci, M. (2013). The mismeasure of machine: Synthetic biology and the trouble with engineering metaphors. *Studies in History and Philosophy of Science Part C: Studies in History and Philosophy of Biological and Biomedical Sciences, 44*(4), 660–668; Bongard, J., & Levin, M. (2021). Living things are not (20th century) machines: Updating mechanism metaphors in light of the modern science of machine behavior. *Frontiers in Ecology and Evolution, 9*, 650726; Esposito, M., & Baravalle, L. (2023). The machine-organism relation revisited. *History and Philosophy of the Life Sciences, 45*(3), 34.

40. Twain, M. (1889). *A Connecticut Yankee in King Arthur's Court.* Charles L. Webster.

41. Thwaites, T. (2011). *The Toaster Project: Or a Heroic Attempt to Build a Simple Electric Appliance from Scratch.* Chronicle Books.

42. Thwaites, T. (2016). *GoatMan: How I Took a Holiday from Being Human.* Chronicle Books.

43. Dartnell, L. (2015). *The Knowledge: How to Rebuild Civilization in the Aftermath of a Cataclysm.* Penguin.

44. It is startling just how important facilities such as the Svalbard Global Seed Vault in the remote Arctic would be to the survival of our species if a global disaster were to occur.

Chapter 5

1. We were repeatedly assured the fistula doesn't bother the cow.

2. For example, Piaget, J. (1952). *The Origins of Intelligence in Children.* International Universities Press; Bruner, J. S., Olver, R. R., & Greenfield, P. M., et al. (1966). *Studies in Cognitive Growth.* Wiley.

3. Varallyay, A., Beller, N., & Subiaul, F. (2023). Generative cultural learning in children and adults: The role of compositionality and generativity in cultural evolution. *Proceedings of the Royal Society B, 290*, 20222418.

4. Liquin, E. G., & Lombrozo, T. (2020). Explanation-seeking curiosity in childhood. *Current Opinion in Behavioral Sciences, 35*, 14–20; Liquin, E. G., & Lombrozo, T. (2020). A functional approach to explanation-seeking curiosity. *Cognitive Psychology, 119*, 101276.

5. Danovitch, J. H., & Mills, C. M. (2018). Understanding when and how explanation promotes exploration. In Saylor, M., & Ganea, P. (Eds.), *Active Learning from Infancy to Childhood* (pp. 95–112). Springer; Danovitch, J. H., Mills, C. M., Sands, K. R.,

& Williams, A. J. (2021). Mind the gap: How incomplete explanations influence children's interest and learning behaviors. *Cognitive Psychology, 130*, 101421.

6. Keil, F. C. (2022). *Wonder: Childhood and the Lifelong Love of Science*. MIT Press.

7. E.g., Schulz, L. (2012). The origins of inquiry: Inductive inference and exploration in early childhood. *Trends in Cognitive Sciences, 16*(7), 382–389.

8. Goddu, M. K., & Gopnik, A. (2024). The development of human causal learning and reasoning. *Nature Reviews Psychology, 3*(5), 319–339.

9. Woo, B. M., Liu, S., & Spelke, E. S. (2022). Infants rationally infer the goals of other people's reaches in the absence of first-person experience with reaching actions. *Developmental Science*, e13453; Liu, S., Brooks, N. B., & Spelke, E. S. (2019). Origins of the concepts cause, cost, and goal in prereaching infants. *Proceedings of the National Academy of Sciences, 116*(36), 17747–17752.

10. Schlottmann, A., Surian, L., & Ray, E. D. (2009). Causal perception of action-and-reaction sequences in 8- to 10-month-olds. *Journal of Experimental Child Psychology, 103*(1), 87–107.

11. Goddu, M. K., & Gopnik, A. (2024). The development of human causal learning and reasoning. *Nature Reviews Psychology, 3*, 319–339.

12. Muentener, P., & Bonawitz, E. B. (2017). The development of causal reasoning. In Waldmann, M. R. (Ed.), *The Oxford Handbook of Causal Reasoning* (pp. 677–698). Oxford University Press.

13. Gelman, S. A. (2023). Looking beyond the obvious. *American Psychologist, 78*(5), 667.

14. Keil, F. C. (1989). *Concepts, Kinds, and Cognitive Development*. MIT Press; Keil, F. C. (1986). The acquisition of natural kind and artifact terms. In Demopoulos, W. (Ed.), *Language Learning and Concept Acquisition: Foundational Issues* (pp. 133–153). Ablex.

15. Newman, G. E., Herrmann, P., Wynn, K., & Keil, F. C. (2008). Biases towards internal features in infants' reasoning about objects. *Cognition, 107*(2), 420–432.

16. Setoh, P., Wu, D., Baillargeon, R., & Gelman, R. (2013). Young infants have biological expectations about animals. *Proceedings of the National Academy of Sciences, 110*(40), 15937–15942; Kominsky, J. F., Li, Y., & Carey, S. (2022). Infants' attributions of insides and animacy in causal interactions. *Cognitive Science, 46*(1), e13087.

17. Wadhams, G. H., & Armitage, J. P. (2004). Making sense of it all: Bacterial chemotaxis. *Nature Reviews Molecular Cell Biology, 5*(12), 1024–1037; Sourjik, V., & Wingreen, N. S. (2012). Responding to chemical gradients: Bacterial chemotaxis. *Current Opinion in Cell Biology, 24*(2), 262–268.

18. Baillargeon, R., Scott, R. M., & Bian, L. (2016). Psychological reasoning in infancy. *Annual Review of Psychology, 67*(1), 159–186; Scott, R. M., Roby, E., & Baillargeon, R. (2022). How sophisticated is infants' theory of mind? In Houdé, O., & Borst, G. (Eds.), *The Cambridge Handbook of Cognitive Development* (p. 242).

19. Caron, A. J., Caron, R. F., & Antell, S. E. (1988). Infant understanding of containment: An affordance perceived or a relationship conceived? *Developmental Psychology, 24*(5), 620; MacLean, D. J., & Schuler, M. (1989). Conceptual development in infancy: The understanding of containment. *Child Development, 60*(5), 1126–1137; Ullman, S., Dorfman, N., & Harari, D. (2019). A model for discovering "containment" relations. *Cognition*, 183, 67–81.

20. Hafri, A., & Firestone, C. (2021). The perception of relations. *Trends in Cognitive Sciences, 25*(6), 475–492.

21. Baillargeon, R., Scott, R. M., & Bian, L. (2016). Psychological reasoning in infancy. *Annual Review of Psychology, 67*, 159–186; Scott, Roby, & Baillargeon. How sophisticated is infants' theory of mind?; Choi, Y., Luo, Y., & Baillargeon, R. (2022). Can 5-month-old infants consider the perspective of a novel eyeless agent? New evidence for early mentalistic reasoning. *Child Development, 93*(2), 571–581.

22. Taborda-Osorio, H., & Cheries, E. W. (2018). Infants' agent individuation: It's what's on the insides that counts. *Cognition, 175*, 11–19; Taborda-Osorio, H., & Cheries, E. W. (2017). Developmental origins of biological explanations: The case of infants' internal property bias. *Psychonomic Bulletin & Review, 24*, 1527–1537; Kominsky, J. F., Shafto, P., & Bonawitz, E. (2021). "There's something inside": Children's intuitions about animate agents. *PLoS One, 16*(5), e0251081. https://doi.org/10.1371/journal.pone.0251081; Baillargeon, Scott, & Bian. Psychological reasoning in infancy.

23. Simons, D. J., & Keil, F. C. (1995). An abstract to concrete shift in the development of biological thought: The insides story. *Cognition, 56*(2), 129–163; Gelman, S. A., & Wellman, H. M. (1991). Insides and essences: Early understandings of the non-obvious. *Cognition, 38*(3), 213–244.

24. Choi, Y., Seok, J., & Luo, Y. (2023). Young infants' expectations about a self-propelled agent's body. *Cognition, 241*, 105629.

25. E.g., Setoh, P., Wu, D., Baillargeon, R., & Gelman, R. (2013). Young infants have biological expectations about animals. *Proceedings of the National Academy of Sciences, 110*(40), 15937–1594; Kominsky, J. F., Li, Y., & Carey, S. (2022). Infants' attributions of insides and animacy in causal interactions. *Cognitive Science, 46*(1), e13087; Taborda-Osorio & Cheries. Developmental origins of biological explanations.

26. Stahl, A. E., & Feigenson, L. (2015). Observing the unexpected enhances infants' learning and exploration. *Science, 348*(6230), 91–94; Stahl, A. E., & Feigenson, L. (2024). Young children distinguish the impossible from the merely improbable.

Proceedings of the National Academy of Sciences, 121(46), e2411297121; Stahl, A. E., & Feigenson, L. (2019). Violations of core knowledge shape early learning. *Topics in Cognitive Science, 11*(1), 136–153; Perez, J., & Feigenson, L. (2022). Violations of expectation trigger infants to search for explanations. *Cognition, 218*, 104942.

27. Stahl, A. E., & Feigenson, L. (2019). Violations of core knowledge shape early learning. *Topics in Cognitive Science, 11*(1), 136–153.

28. Keil. *Wonder.*

29. Stahl & Feigenson. Young children distinguish the impossible.

30. Cesana-Arlotti, N., Martín, A., Téglás, E., Vorobyova, L., Cetnarski, R., & Bonatti, L. L. (2018). Precursors of logical reasoning in preverbal human infants. *Science, 359*(6381), 1263–1266; Cesana-Arlotti, N., Kovács, Á. M., & Téglás, E. (2020). Infants recruit logic to learn about the social world. *Nature Communications, 11*(1), 5999.

31. Newman, G. E., & Keil, F. C. (2008). Where is the essence? Developmental shifts in children's beliefs about internal features. *Child Development, 79*(5), 1344–1356.

32. Zhang, S., She, J. S., Gerstenberg, T., & Rose, D. (2023). You are what you're for: Essentialist categorization in large language models. *Proceedings of the Annual Meeting of the Cognitive Science Society, 45*(45).

33. Sobel, D. M., Yoachim, C. M., Gopnik, A., Meltzoff, A. N., & Blumenthal, E. J. (2007). The blicket within: Preschoolers' inferences about insides and causes. *Journal of Cognition and Development, 8*(2), 159–182.

34. Kominsky, J. F., Zamm, A. P., & Keil, F. C. (2018). Knowing when help is needed: A developing sense of causal complexity. *Cognitive Science, 42*(2), 491–523.

35. E.g., Ulrich, H., Kock-Schoppenhauer, A. K., Deppenwiese, N., Gött, R., Kern, J., Lablans, M., . . . & Ingenerf, J. (2022). Understanding the nature of metadata: Systematic review. *Journal of Medical Internet Research, 24*(1), e25440.

36. Erb, C. D., Buchanan, D. W., & Sobel, D. M. (2013). Children's developing understanding of the relation between variable causal efficacy and mechanistic complexity. *Cognition, 129*(3), 494–500. https://doi.org/10.1016/j.cognition.2013.08.002

37. Bullock, M., Gelman, R., & Baillargeon, R. (1982). *The Development of Causal Reasoning.* Bullock, M., Gelman, R., Baillargeon, R. (1982). The development of causal reasoning. In Friedman, W. J. (Ed.), *The Developmental Psychology of Time* (pp. 209–254). Academic Press.

38. Ahl, R. E., & Keil, F. C. (2017). Diverse effects, complex causes: Children use information about machines' functional diversity to infer internal complexity. *Child Development, 88*, 828–845; Ahl, R. E., DeAngelis, E., & Keil, F. C. (2022). "I know it's complicated": Children detect relevant information about object complexity. *Journal of Experimental Child Psychology, 222*, 105465; Ahl, R. E., Amir, D., & Keil, F. C.

(2020). The world within: Children are sensitive to internal complexity cues. *Journal of Experimental Child Psychology, 200*, 104932.

39. Keil, F. C., & Lockhart, K. L. (2021). Beyond cause: The development of clockwork cognition. *Current Directions in Psychological Science, 30*(2), 167–173.

40. Lockhart, K. L., Chuey, A., Kerr, S., & Keil, F. C. (2019). The privileged status of knowing mechanistic information: An early epistemic bias. *Child Development, 90*(5), 1772–1788.

41. Sands, K. R., Monroe, A. J., & Mills, C. M. (2023). "How do fish breathe underwater?": Young children's ability to evaluate and remember different types of explanations regarding biological phenomena. *Cognitive Development, 66*, 101330; see also Mills, C. M., Sands, K. R., Rowles, S. P., & Campbell, I. L. (2019). "I want to know more!": Children are sensitive to explanation quality when exploring new information. *Cognitive Science, 43*(1), e12706; Attisano, E., Nancekivell, S. E., & Denison, S. (2021). Components and mechanisms: How children talk about machines in museum exhibits. *Frontiers in Psychology, 12*, 636601.

42. Hegarty, M. (2004). Mechanical reasoning by mental simulation. *Trends in Cognitive Sciences, 8*(6), 280–285.

43. Gilligan-Lee, K. A., Hawes, Z. C., & Mix, K. S. (2022). Spatial thinking as the missing piece in mathematics curricula. *npj Science of Learning, 7*(1), 10.

44. Judd, N., & Klingberg, T. (2021). Training spatial cognition enhances mathematical learning in a randomized study of 17,000 children. *Nature Human Behaviour, 5*(11), 1548–1554; Cheng, Y. L., & Mix, K. S. (2014). Spatial training improves children's mathematics ability. *Journal of Cognition and Development, 15*(1), 2–11; Newcombe, N. S. (2010). Picture this: Increasing math and science learning by improving spatial thinking. *American Educator, 34*(2), 29–35; Hawes, Z. C., Gilligan-Lee, K. A., & Mix, K. S. (2022). Effects of spatial training on mathematics performance: A meta-analysis. *Developmental Psychology, 58*(1), 112–137.

45. Casasola, M., Wei, W. S., Suh, D. D., Donskoy, P., & Ransom, A. (2020). Children's exposure to spatial language promotes their spatial thinking. *Journal of Experimental Psychology: General, 149*(6), 111–116.

46. Epstein, R. A., Patai, E. Z., Julian, J. B., & Spiers, H. J. (2017). The cognitive map in humans: Spatial navigation and beyond. *Nature Neuroscience, 20*(11), 1504.

47. Snow, C. P. (2012). *The Two Cultures*. Cambridge University Press.

48. Sen, P. (2021). *Einstein's Fridge: How the Difference Between Hot and Cold Explains the Universe*. Simon and Schuster.

49. Newman, G. E., Keil, F. C., Kuhlmeier, V. A., & Wynn, K. (2010). Early understandings of the link between agents and order. *Proceedings of the National Academy of Sciences, 107*(40), 17140–17145.

50. Woodward, A. L. (1998). Infants selectively encode the goal object of an actor's reach. *Cognition, 69*(1), 1–34.

51. Ma, L., & Xu, F. (2013). Preverbal infants infer intentional agents from the perception of regularity. *Developmental Psychology, 49*(7), 1330–1337.

52. Gergely, G., Nádasdy, Z., Csibra, G., & Bíró, S. (1995). Taking the intentional stance at 12 months of age. *Cognition, 56*(2), 165–193.

53. Jara-Ettinger, J., Gweon, H., Schulz, L. E., & Tenenbaum, J. B. (2016). The naïve utility calculus: Computational principles underlying commonsense psychology. *Trends in Cognitive Sciences, 20*(8), 589–604.

54. Jara-Ettinger, J., Schulz, L. E., & Tenenbaum, J. B. (2020). The naive utility calculus as a unified, quantitative framework for action understanding. *Cognitive Psychology*, 123, 101334.

55. Tversky, A., & Kahneman, D. (1974). Judgment under uncertainty: Heuristics and biases: Biases in judgments reveal some heuristics of thinking under uncertainty. *Science, 185*(4157), 1124–1131; Gilovich, T., Griffin, D., & Kahneman, D. (Eds.). (2002). *Heuristics and Biases: The Psychology of Intuitive Judgment.* Cambridge University Press.

56. Liu, S., Ullman, T. D., Tenenbaum, J. B., & Spelke, E. S. (2017). Ten-month-old infants infer the value of goals from the costs of actions. *Science, 358*(6366), 1038–1041.

57. Liu, S., & Spelke, E. S. (2017). Six-month-old infants expect agents to minimize the cost of their actions. *Cognition, 160*, 35–42.

58. Leonard, J. A., Lee, Y., & Schulz, L. E. (2017). Infants make more attempts to achieve a goal when they see adults persist. *Science, 357*(6357), 1290–1294; Leonard, J. A., Martinez, D. N., Dashineau, S. C., Park, A. T., & Mackey, A. P. (2021). Children persist less when adults take over. *Child Development, 92*(4), 1325–1336.

59. Sehl, C. G., Denison, S., & Friedman, O. (2022). Using efficiency to infer the quality of machines. In *Proceedings of the Annual Meeting of the Cognitive Science Society, 44*(44).

60. Dreyfuss, H. (1959). Designing for people. *Design, 61*(2), 80–83.

61. Sobel, D. (2007). *Longitude: The True Story of a Lone Genius Who Solved the Greatest Scientific Problem of His Time.* Bloomsbury.

62. Thormann, K. M., Beta, C., & Kühn, M. J. (2022). Wrapped up: The motility of polarly flagellated bacteria. *Annual Review of Microbiology, 76*, 349–367; Nicolau Jr, D. V., Armitage, J. P., & Maini, P. K. (2009). Directional persistence and the optimality of run-and-tumble chemotaxis. *Computational Biology and Chemistry, 33*(4), 269–274.

63. Wiener, J. M., Ehbauer, N. N., & Mallot, H. A. (2009). Planning paths to multiple targets: Memory involvement and planning heuristics in spatial problem solving. *Psychological Research PRPF, 73*(5), 644–658; Yi, S. K. M., Steyvers, M., Lee, M. D., & Dry, M. J. (2012). The wisdom of the crowd in combinatorial problems. *Cognitive Science, 36*(3), 452–470.

64. Smith-Ferguson, J., & Beekman, M. (2020). Who needs a brain? Slime moulds, behavioural ecology and minimal cognition. *Adaptive Behavior, 28*(6), 465–478.

Chapter 6

1. Harris, P. L. (2012). *Trusting What You're Told: How Children Learn from Others.* Harvard University Press; Harris, P. L., Koenig, M. A., Corriveau, K. H., & Jaswal, V. K. (2018). Cognitive foundations of learning from testimony. *Annual Review of Psychology, 69*, 251–273; Gelman, S. A. (2009). Learning from others: Children's construction of concepts. *Annual Review of Psychology, 60*, 115–140; Gweon, H. (2021). Inferential social learning: Cognitive foundations of human social learning and teaching. *Trends in Cognitive Sciences, 25*(10), 896–910.

2. Leslie, A. M. (1987). Pretense and representation: The origins of "theory of mind." *Psychological Review, 94*(4), 412; Harris, P. L., Kavanaugh, R. D., Wellman, H. M., & Hickling, A. K. (1993). Young children's understanding of pretense. *Monographs of the Society for Research in Child Development, 58*(1), i–107; Harris, P. L. (2022). Young children share imagined possibilities: evidence for an early-emerging human competence. *Philosophical Transactions of the Royal Society B, 377*(1866), 20220022; Harris, P. L. (2024). Somewhere, somehow: Young children imagine possibilities and ask about them. *Annual Review of Developmental Psychology, 6*(1), 185–201.

3. Callanan, M. A., Legare, C. H., Sobel, D. M., Jaeger, G. J., Letourneau, S., McHugh, S. R., . . . & Watson, J. (2020). Exploration, explanation, and parent–child interaction in museums. *Monographs of the Society for Research in Child Development, 85*(1), 7–137.

4. Gripshover, S. J., & Markman, E. M. (2013). Teaching young children a theory of nutrition: Conceptual change and the potential for increased vegetable consumption. *Psychological Science, 24*(8), 1541–1553; Smith, E. L., Blakeslee, T. D., & Anderson, C. W. (1993). Teaching strategies associated with conceptual change learning in science. *Journal of Research in Science Teaching, 30*(2), 111–126; Evans, E. M. (2008). Conceptual change and evolutionary biology: A developmental analysis. In Vosniadou, E. (Ed.), *International Handbook of Research on Conceptual Change* (pp. 263–294).

5. Mills, C. M. (2013). Knowing when to doubt: Developing a critical stance when learning from others. *Developmental Psychology, 49*(3), 404–418; Williams, A. J., & Danovitch, J. H. (2023). The role of accuracy in children's judgments of experts' knowledge. *Child Development, 95*, 128–143.

6. Keil, F. C., Stein, C., Webb, L., Billings, V. D., & Rozenblit, L. (2008). Discerning the division of cognitive labor: An emerging understanding of how knowledge is clustered in other minds. *Cognitive Science, 32*(2), 259–300; Lutz, D. J., & Keil, F. C. (2002). Early understanding of the division of cognitive labor. *Child Development, 73*(4), 1073–1084; Danovitch, J. H. (2020). Children's selective information sharing based on the recipient's role. *Journal of Genetic Psychology, 181*(2–3), 68–77.

7. Danovitch, J. H., & Keil, F. C. (2004). Should you ask a fisherman or a biologist? Developmental shifts in ways of clustering knowledge. *Child Development, 75*(3), 918–931.

8. Thagard, P. R. (1978). Why astrology is a pseudoscience. In *PSA: Proceedings of the Biennial Meeting of the Philosophy of Science Association* (pp. 223–234). Cambridge University Press; Blancke, S., Edis, T., Braeckman, J., Hansson, S. O., Landrum, A. R., & Shtulman, A. (2022). The psychology of pseudoscience. *Frontiers in Psychology, 13*, 935645.

9. Yousif, S. R., Aboody, R., & Keil, F. C. (2019). The illusion of consensus: A failure to distinguish between true and false consensus. *Psychological Science, 30*(8), 1195–1204.

10. Robb, A. (2017, November 16). Anatomy of a fake news scandal. *Rolling Stone.*

11. Aboody, R., Yousif, S. R., Sheskin, M., & Keil, F. C. (2022). Says who? Children consider informants' sources when deciding whom to believe. *Journal of Experimental Psychology: General, 151*(10), 2481.

12. E.g., Rowse, A. E. (1992). How to build support for a war. *Columbia Journalism Review, 31*(3), 28–30. Oddo, J. (2018). *The Discourse of Propaganda: Case Studies from the Persian Gulf War and the War on Terror.* Penn State Press; Jones, M. O. (2022). *Digital Authoritarianism in the Middle East: Deception, Disinformation and Social Media.* Oxford University Press.

13. Mercier, H., & Sperber, D. (2011). Why do humans reason? Arguments for an argumentative theory. *Behavioral and Brain Sciences, 34*, 57–111; Li, L., & Tomasello, M. (2022). Disagreement, justification, and equitable moral judgments: A brief training study. *Journal of Experimental Child Psychology, 223*, 105494.

14. Heyes, C., Bang, D., Shea, N., Frith, C. D., & Fleming, S. M. (2020). Knowing ourselves together: The cultural origins of metacognition. *Trends in Cognitive Sciences, 24*, 349–362.

15. Meyer, M. M., Proft, M., Engelmann, J., & Rakoczy, H. (2025). Young children demonstrate improved metacognitive competence in social contexts. PsyArXiv. https://dx.doi.org/10.31234/osf.io/9qz6b

16. Dunham, Y. (2018). Mere membership. *Trends in Cognitive Sciences, 22*, 780–793.

17. Kominsky, J. F., Langthorne, P., & Keil, F. C. (2016). The better part of not knowing: Virtuous ignorance. *Developmental Psychology, 52*, 31.

18. Dunning, D. (2011). The Dunning–Kruger effect: On being ignorant of one's own ignorance. In Olson, J. M., & Zanna, M. P. (Eds.), *Advances in Experimental Social Psychology* (Vol. 44, pp. 247–296). Academic Press.

19. Darwin, C. (1871). Introduction, *The Descent of Man* (p. 3). John Murray.

20. From Maxwell's Inaugural Address as the first Cavendish Professor of Experimental Physics Laboratory at Cambridge University in 1871: Maxwell, J. C. (1871). Introductory Lecture of Experimental Physics. Macmillon; Russell, B. (1933). The triumph of stupidity. In Russell, B. (1998). *Mortals and Others* (Vol. 2: American Essays 1931–1935) (H. Ruja, Ed.). Routledge.

21. Mills, C. M., Danovitch, J. H., & Quintero, N.B. (2025). Does intellectual humility transmit intergenerationally? Examining relations between parent and child measures. *Journal of Experimental Psychology: General, 154*, 2272–2284.

22. Inhelder, B., & Piaget, J. (1958). *The Growth of Logical Thinking: From Childhood to Adolescence*. Basic Books.

23. Kominsky, J. F., Gerstenberg, T., Pelz, M., Sheskin, M., Singmann, H., Schulz, L., & Keil, F. C. (2021). The trajectory of counterfactual simulation in development. *Developmental Psychology, 57*(2), 253–268.

24. Chi, M. T., Feltovich, P. J., & Glaser, R. (1981). Categorization and representation of physics problems by experts and novices. *Cognitive Science, 5*(2), 121–152.

25. Chi, M. T. (1977). Age differences in memory span. *Journal of Experimental Child Psychology, 23*(2), 266–281; Schneider, W., Gruber, H., Gold, A., & Opwis, K. (1993). Chess expertise and memory for chess positions in children and adults. *Journal of Experimental Child Psychology, 56*(3), 328–349; Chi, M. T. H. (1978). Knowledge structures and memory development. In R. Siegler (Ed.), *Children's Thinking: What Develops* (pp. 73–96). Erlbaum.

26. Harris, P. L., German, T., & Mills, P. (1996). Children's use of counterfactual thinking in causal reasoning. *Cognition, 61*(3), 233–259; Wong, A., Cordes, S., Harris, P. L., & Chernyak, N. (2023). Being nice by choice: The effect of counterfactual reasoning on children's social evaluations. *Developmental Science, 26*, e13394; Harris, P. L., & Leevers, H. J. (2014). Reasoning from false premises. In Mitchell, P., & Riggs, K. (Eds.), *Children's Reasoning and the Mind* (pp. 67–86). Psychology Press.

27. Kretch, K. S., Franchak, J. M., & Adolph, K. E. (2014). Crawling and walking infants see the world differently. *Child Development, 85*(4), 1503–1518.

28. Shtulman, A. (2023). *Learning to Imagine: The Science of Discovering New Possibilities*. Harvard University Press.

29. Gill, M., & Keil, F. C. (2026). Forseeability and liability (unpublished ms).

30. Beck, S. R., Apperly, I. A., Chappell, J., Guthrie, C., & Cutting, N. (2011). Making tools isn't child's play. *Cognition, 119*, 301–306; German, T. P., & Defeyter, M. A. (2000). Immunity to functional fixedness in young children. *Psychonomic Bulletin & Review, 7*(4), 707–712.

31. Cutting, N., Apperly, I. A., & Beck, S. R. (2011). Why do children lack the flexibility to innovate tools? *Journal of Experimental Child Psychology, 109*, 497–551; Weir, A. A. S., Chappell, J., & Kacelnik, A. (2002). Shaping of hooks in New Caledonian crows. *Science, 297*(5583); Cutting, N., Apperly, I. A., Chappell, J., & Beck, S. R. (2014). The puzzling difficulty of tool innovation: Why can't children piece their knowledge together? *Journal of Experimental Child Psychology, 125*, 110–117; Neldner, K., Redshaw, J., Murphy, S., Tomaselli, K., Davis, J., Dixson, B., & Nielsen, M. (2019). Creation across culture: Children's tool innovation is influenced by cultural and developmental factors. *Developmental Psychology, 55*(4), 877–889; Rawlings, B., & Legare, C. H. (2021). Toddlers, tools, and tech: The cognitive ontogenesis of innovation. *Trends in Cognitive Sciences, 25*(1), 81–92.

32. Shtulman, A. (2023). *Learning to Imagine: The Science of Discovering New Possibilities*. Harvard University Press.

33. Lyons, D. E., Young, A. G., & Keil, F. C. (2007). The hidden structure of overimitation. *Proceedings of the National Academy of Sciences, 104*(50), 19751–19756.

34. Horner, V., & Whiten, A. (2005). Causal knowledge and imitation/emulation switching in chimpanzees (*Pan troglodytes*) and children (*Homo sapiens*). *Animal Cognition, 8*(3), 164–181.

35. Lyons, D. E., Damrosch, D. H., Lin, J. K., Macris, D. M., & Keil, F. C. (2011). The scope and limits of overimitation in the transmission of artefact culture. *Philosophical Transactions of the Royal Society B: Biological Sciences, 366*(1567), 1158–1167.

36. Johnston, A. M., Holden, P. C., & Santos, L. R. (2017). Exploring the evolutionary origins of overimitation: A comparison across domesticated and non-domesticated canids. *Developmental Science, 20*(4), e12460.

37. Henrich, N., & Henrich, J. P. (2007). *Why Humans Cooperate: A Cultural and Evolutionary Explanation*. Oxford University Press.

38. Danovitch, J. H., & Keil, F. C. (2004). Should you ask a fisherman or a biologist? Developmental shifts in ways of clustering knowledge. *Child Development, 75*(3), 918–931.

39. Hammack, B. (2023). *The Things We Make: The Unknown History of Invention from Cathedrals to Soda Cans*. Sourcebooks; Agrawal, R. (2023). *Nuts and Bolts: Seven Small Inventions That Changed the World in a Big Way*. W. W. Norton.

40. See for example Dougherty, D. (2012). The maker movement. *Innovations: Technology, Governance, Globalization, 7*(3), 11–14.

41. Burdett, E. R., & Ronfard, S. (2023). Tinkering to innovation: How children refine tools over multiple attempts. *Developmental Psychology, 59*(6), 1006; Ockerby, Z., Redshaw, J., & Suddendorf, T. (2025). A key to innovation: When do children begin to recognize and manufacture solutions to future problems? *Developmental Psychology*. Advance online publication. https://doi.org/10.1037/dev0001926

42. Neldner, K., Redshaw, J., Murphy, S., Tomaselli, K., Davis, J., Dixson, B., & Nielsen, M. (2019). Creation across culture: Children's tool innovation is influenced by cultural and developmental factors. *Developmental Psychology, 55*(4), 877–889.

43. E.g., Gönül, G., Takmaz, E., & Hohenberger, A. (2021). Preschool children's use of perceptual-motor knowledge and hierarchical representational skills for tool making. *Acta Psychologica, 220*, 103415.

44. Shan, J., & Wang, W. (2023). Making and sharing in asynchronous discussion: Exploring the collaboration process in online maker community. *Interactive Learning Environments, 31*(5), 3003–3017; Vossoughi, S., & Bevan, B. (2014). Making and tinkering: A review of the literature. *National Research Council Committee on Out of School Time STEM*, 67, 1–55.

45. Gibbs, N. (2009). The growing backlash against overparenting. *Time*, 1–7; Skenazy, L. (2021). *Free-Range Kids: How Parents and Teachers Can Let Go and Let Grow.* John Wiley & Sons; Gray, P., Lancy, D. F., & Bjorklund, D. F. (2023). Decline in independent activity as a cause of decline in children's mental well-being: Summary of the evidence. *Journal of Pediatrics, 260*; Boyatzis, C. J., King, K. B., Doak, M., & Zajac, L. A. (2025). "It's very much how you define dangerous": The cultural construction of mothers' beliefs about risk in Denmark and the United States. *Acta Psychologica, 255*, 104917.

Chapter 7

1. Paul, P. (2023, October 12). To prepare kids for the future, bring back shop and home ec. *New York Times*. https://www.nytimes.com/2023/10/12/opinion/school-manual-labor-shop.html

2. E.g., Bennett, J. (2014, August 26). Majority of teens say they don't know how to change a tire: Survey finds 44% don't know how to check tread depth, 32% don't know how to check tire pressure. *Wall Street Journal.*

3. Hembree, R., & Dessart, D. J. (1986). Effects of hand-held calculators in precollege mathematics education: A meta-analysis. *Journal for Research in Mathematics Education, 17*(2), 83–89; Ellington, A. J. (2003). A meta-analysis of the effects of calculators on students' achievement and attitude levels in precollege mathematics

classes. *Journal for Research in Mathematics Education, 34*(5), 433–463; Rittle-Johnson, B., & Kmicikewycz, A. O. (2008). When generating answers benefits arithmetic skill: The importance of prior knowledge. *Journal of Experimental Child Psychology, 101*(1), 75–81.

4. Iyengar, S. (2024). Federal data on reading for pleasure: All signs show a slump. *National Endowment for the Arts Blog*. https://www.arts.gov/stories/blog/2024/federal-data-reading-pleasure-all-signs-show-slump

5. Silcoff, M. (2024, September 2). I paid my child $100 to read a book. *New York Times*.

6. Bauerlein, M. (2022). *The Dumbest Generation Grows Up: From Stupefied Youth to Dangerous Adults*. Simon and Schuster; Bauerlein, M. (2009). *The Dumbest Generation: How the Digital Age Stupefies Young Americans and Jeopardizes Our Future*. Jeremy P. Tarcher/Penguin.

7. Carr, N. (2020). *The Shallows: What the Internet Is Doing to Our Brains*. W. W. Norton; see also Gerlich, M. (2025). AI tools in society: Impacts on cognitive offloading and the future of critical thinking. *Societies, 15*(1), 6.

8. Winchester, S. (2023). *Knowing What We Know: The Transmission of Knowledge from Ancient Wisdom to Modern Magic*. Harper Perennial.

9. Carr, N. (2025). *Superbloom: How Technologies of Connection Tear Us Apart*. W. W. Norton.

10. Twenge, J. M. (2023). *Generations: The Real Differences Between Gen Z, Millennials, Gen X, Boomers, and Silents—and What They Mean for America's Future*. Simon and Schuster.

11. See also Twenge, J., & Blanchflower, D. G. (2025). *Declining Life Satisfaction and Happiness Among Young Adults in Six English-speaking Countries* (No. w33490). National Bureau of Economic Research.

12. Prasad, S., Souabni, S. A., Anugwom, G., Aneni, K., Anand, A., Urhi, A., . . . & Oladunjoye, F. (2023). Anxiety and depression amongst youth as adverse effects of using social media: A review. *Annals of Medicine and Surgery, 85*(8), 3974–3981.

13. E.g., Hunt, M. G., Marx, R., Lipson, C., & Young, J. (2018). No more FOMO: Limiting social media decreases loneliness and depression. *Journal of Social and Clinical Psychology, 37*(10), 751–768.

14. Haidt, J. (2024). *The Anxious Generation: How the Great Rewiring of Childhood Is Causing an Epidemic of Mental Illness*. Penguin.

15. Uncapher, M. R., & Wagner, A. D. (2018). Minds and brains of media multitaskers: Current findings and future directions. *Proceedings of the National Academy of Sciences, 115*(40), 9889–9896; Baumgartner, S. E., Parry, D. A., Beyens, I., Wiradhany, W.,

Uncapher, M., Wagner, A. D., & Bavelier, D. (2024). The short- and long-term effects of digital media use on attention. In Christakis, D. A., & Hale, L. (Eds.), *Handbook of Children and Screens: Digital Media, Development, and Well-Being from Birth Through Adolescence* (pp. 31–37). Springer Nature; Thiele, L. P. (2025). Deskilling: The atrophy of cognitive and social aptitudes. In *Human Agency, Artificial Intelligence, and the Attention Economy: The Case for Digital Distancing* (pp. 113–152). Springer Nature; Liao, S., Yu, L., Kruger, J. L., & Reichle, E. D. (2024). Dynamic reading in a digital age: New insights on cognition. *Trends in Cognitive Sciences, 28*(1), 43–55.

16. Johnson, S. (2005). *Everything Bad Is Good for You: How Today's Popular Culture Is Actually Making Us Smarter* (p. 11). Penguin.

17. Dworak, E. M., Revelle, W., & Condon, D. M. (2023). Looking for Flynn effects in a recent online US adult sample: Examining shifts within the SAPA Project. *Intelligence, 98*, 101734. Nordmo, M., Norrøne, T. N., & Lang-Ree, O. C. (2025). Reevaluating the Flynn effect, and the reversal: Temporal trends and measurement invariance in Norwegian armed forces intelligence scores. *Intelligence, 110*, 101909; Goldstein, T. (2023, October 11). ACT reports record low scores as admissions landscape shifts. *New York Times*.

18. Toppo, G. (2015). *The Game Believes in You: How Digital Play Can Make Our Kids Smarter*. St. Martin's Press; McGoniga, J. (2011). *Reality Is Broken: Why Games Make Us Better and How They Can Change the World*. Penguin.

19. Haidt. *The Anxious Generation*.

20. Lienhard, J. H. (2003). *Inventing Modern: Growing Up with X-Rays, Skyscrapers, and Tailfins*. Oxford University Press.

21. Keil, F. C. (2006). Explanation and understanding. *Annual Review of Psychology, 57*(1), 227–254.

22. Thagard, P. (2021). Naturalizing logic: How knowledge of mechanisms enhances inductive inference. *Philosophies, 6*(2), 52.

23. Zemla, J. C., Sloman, S., Bechlivanidis, C., & Lagnado, D. A. (2023). Not so simple! Causal mechanisms increase preference for complex explanations. *Cognition, 239*, 105551.

24. E.g., Chongchitpaisan, P., & Sudsawat, S. (2022). A review on screw conveyors for bulk materials in various applications. *Engineering and Technology Horizons, 39*(2), 1–12.

25. Sweeney, H. L., & Hammers, D. W. (2018). Muscle contraction. *Cold Spring Harbor Perspectives in Biology, 10*(2), a023200; Willingham, T. B., Kim, Y., Lindberg, E., Bleck, C. K., & Glancy, B. (2020). The unified myofibrillar matrix for force generation in muscle. *Nature Communications, 11*(1), 3722.

26. E.g., Howell, E. L., & Brossard, D. (2021). (Mis)informed about what? What it means to be a science-literate citizen in a digital world. *Proceedings of the National Academy of Sciences, 118*(15), e1912436117; Cacciatore, M. A. (2021). Misinformation and public opinion of science and health: Approaches, findings, and future directions. *Proceedings of the National Academy of Sciences, 118*(15), e1912437117; Swire-Thompson, B., Miklaucic, N., Wihbey, J. P., Lazer, D., & DeGutis, J. (2022). The backfire effect after correcting misinformation is strongly associated with reliability. *Journal of Experimental Psychology: General, 151*(7), 1655–1665; Sharon, A. J., & Baram-Tsabari, A. (2020). Can science literacy help individuals identify misinformation in everyday life? *Science Education, 104*(5), 873–894.

27. Greifeneder, R., Jaffe, M., Newman, E., & Schwarz, N. (2021). *The Psychology of Fake News: Accepting, Sharing, and Correcting Misinformation*. Routledge; Sinatra, G. M., & Lombardi, D. (2020). Evaluating sources of scientific evidence and claims in the post-truth era may require reappraising plausibility judgments. *Educational Psychologist, 55*(3), 120–131.

28. Prike, T., & Ecker, U. K. (2023). Effective correction of misinformation. *Current Opinion in Psychology, 54*, 101712.

29. Jamieson, K. H. (2021). How to debunk misinformation about COVID, vaccines and masks. *Scientific American, 324*(4), 44–51.

30. Pinker, S. (1997). *How the Mind Works*. W. W. Norton; Sterelny, K. (2012). *The Evolved Apprentice*. MIT Press.

31. Roozenbeek, J., Culloty, E., & Suiter, J. (2023). Countering misinformation. *European Psychologist, 28*(3); Amazeen, M. A., & Krishna, A. (2024). Refuting misinformation: Examining theoretical underpinnings of refutational interventions. *Current Opinion in Psychology, 56*, 101774; Swire-Thompson, B., DeGutis, J., & Lazer, D. (2020). Searching for the backfire effect: Measurement and design considerations. *Journal of Applied Research in Memory and Cognition, 9*(3), 286–299.

32. Nyhan, B., Reifler, J., Richey, S., & Freed, G. L. (2014). Effective messages in vaccine promotion: A randomized trial. *Pediatrics, 133*(4), e835–e842.

33. Nyhan, B. (2021). Why the backfire effect does not explain the durability of political misperceptions. *Proceedings of the National Academy of Sciences, 118*(15), e1912440117.

34. For a truly inappropriate use of sources see Russell, E., & Russell, J. (2023). The Porter letter: The link between long-term opioids and addiction. In Russell, J., & Skolnik, N. S. (Eds.), *Top Articles in Primary Care* (pp. 269–271). Springer; Becker, W. C., & Fiellin, D. A. (2017). Limited evidence, faulty reasoning, and potential for a global opioid crisis. *BMJ*, 358; Porter, J. (1980). Addiction rare in patients treated with narcotics. *New England Journal of Medicine, 302*(2), 123.

35. Goldwater, M. B., Gentner, D., LaDue, N. D., & Libarkin, J. C. (2021). Analogy generation in science experts and novices. *Cognitive Science, 45*(9), e13036; Holyoak, K. J. (2025). *The Human Edge: Analogy and the Roots of Creative Intelligence*. MIT Press.

36. Tversky, A. (1977). Features of similarity. *Psychological Review, 84*(4), 327–352; Bowdle, B. F., & Gentner, D. (1997). Informativity and asymmetry in comparisons. *Cognitive Psychology, 34*(3), 244–286; Mo, F., & Gentner, D. (2024). Exploring analogical asymmetry. In *Proceedings of the Annual Meeting of the Cognitive Science Society* (Vol. 46, pp. 4027–4032).

37. Liu, S., Outa, J., & Karakose-Akbiyik, S. (2024). Naive psychology depends on naive physics. OSF. https://doi.org/10.31234/osf.io/u6xdz

38. Johnson, S. G., Bilovich, A., & Tuckett, D. (2023). Conviction narrative theory: A theory of choice under radical uncertainty. *Behavioral and Brain Sciences, 46*, e82.

39. Mangalam, M., Fragaszy, D. M., Wagman, J. B., Day, B. M., Kelty-Stephen, D. G., Bongers, R. M., . . . & Osiurak, F. (2022). On the psychological origins of tool use. *Neuroscience & Biobehavioral Reviews, 134*, 104521.

Chapter 8

1. Lewis, D. (2008). *Convention: A Philosophical Study*. John Wiley & Sons.

2. E.g., Larsen, J., Urry, J., & Axhausen, K. (2008). Coordinating face-to-face meetings in mobile network societies. *Information, Communication & Society, 11*(5), 640–658; Thulin, E., & Vilhelmson, B. (2007). Mobiles everywhere: Youth, the mobile phone, and changes in everyday practice. *Young, 15*(3), 235–253; Ling, R., & Lai, C. H. (2016). Microcoordination 2.0: Social coordination in the age of smartphones and messaging apps. *Journal of Communication, 66*(5), 834–856.

3. Dahmani, L., & Bohbot, V. D. (2020). Habitual use of GPS negatively impacts spatial memory during self-guided navigation. *Scientific Reports, 10*(1), 6310.

4. Oral, U. (2023). Netiquette: Fundamentals of etiquette in digital communication. *European Journal of Theoretical and Applied Sciences, 1*(5), 833–847: Katz, R., Ogilvie, S., Shaw, J., & Woodhead, L. (2021). Gen Z, explained: The art of living in a digital age. In *Gen Z, Explained* (pp. 189–204). University of Chicago Press; Peredy, Z., Vigh, L., Wei, Q., & Jiang, M. (2024). Analysing Generation Z communication attitudes, values and norms. *Acta Periodica (Edutus)*, 30, 4–19.

5. Bindley, K. (2024, June 2). Don't you dare call me without texting first. *Wall Street Journal*. See also Jolly, J. (2023, October 6). Punctuation is "judgey"? Text before calling? How proper cellphone etiquette has changed. *USA Today*.

6. Willams, Z. (2023, July 25). Yes, I leave texts unanswered—but don't you dare do the same to me. *The Guardian*. See also Janssen, D., & Carradini, S. (2021). Generation

Z workplace communication habits and expectations. *IEEE Transactions on Professional Communication, 64*(2), 137–153; Holton, T., & Fraser, B. (2015). Generation Z and technology. *Defence Research and Development Canada, 11*(3), 185–186; Gabrielova, K., & Buchko, A. A. (2021). Here comes Generation Z: Millennials as managers. *Business Horizons, 64*(4), 489–499; Bai, S. (2023). The Gen Z revolt. *Maclean's, 136*(2), 20.

7. Lockhart, K. L., Abrahams, B., & Osherson, D. N. (1977). Children's understanding of uniformity in the environment. *Child Development, 48*(4), 1521–1531.

8. Lockhart, K. L. (1981). The development of knowledge about uniformities in the environment: A comparative analysis of the child's understanding of social, moral, and physical rules. Ph.D. dissertation, University of Pennsylvania.

9. Turiel, E. (1983). *The Development of Social Knowledge: Morality and Convention.* Cambridge University Press.

10. Wegner, D. M., Giuliano, T., & Hertel, P. (1985). Cognitive interdependence in close relationships. In W. J. Ickes (Ed.), *Compatible and Incompatible Relationships* (pp. 253–276). Springer-Verlag; Wegner, D. M. (1986). Transactive memory: A contemporary analysis of the group mind. In Mullen, B., & Goethals, G. R. (Eds.), *Theories of Group Behavior* (pp. 185–205). Springer-Verlag.

11. Hollingshead, A. (2001). Cognitive interdependence and convergent expectations in transactive memory. *Journal of Personality and Social Psychology, 81*(6), 1080–1089.

12. O'Toole, J., Ciuchta, M. P., Neville, F., & Lahiri, A. (2023). Transactive memory systems, temporary teams, and conflict: Innovativeness during a hackathon. *Journal of Management, 49*(5), 1633–1661; Jarvenpaa, S. L., & Keating, E. (2024). Fluid teams in the metaverse: Exploring the (un)familiar. *Frontiers in Psychology, 14*, 1323586.

13. de Ridder, J. (2024). Online illusions of understanding. *Social Epistemology, 38*(6), 727–742; He, G., Kuiper, L., & Gadiraju, U. (2023). Knowing about knowing: An illusion of human competence can hinder appropriate reliance on AI systems. In *Proceedings of the 2023 CHI Conference on Human Factors in Computing Systems* 113, 1–18. https://doi.org/10.1145/3544548.3581025

14. Sterelny, K. (2012). *The Evolved Apprentice.* MIT Press.

15. E.g., Smith, A. (1904; orig. 1776). *An Inquiry into the Nature and Causes of the Wealth of Nations* by Adam Smith, edited with an Introduction, Notes, Marginal Summary and an Enlarged Index by Edwin Cannan. 2 vols. Methuen.

16. Good, M. J. D. (2023). *American Medicine: The Quest for Competence.* University of California Press.

17. Donohoe, M. T. (1998). Comparing generalist and specialty care: Discrepancies, deficiencies, and excesses. *Archives of internal medicine, 158*(15), 1596–1608;

Bodenheimer, T. (2022). Revitalizing primary care, part 1: Root causes of primary care's problems. *Annals of Family Medicine, 20*(5), 464–468. Bodenheimer, T. (2022). Revitalizing primary care, part 2: Hopes for the future. *Annals of Family Medicine, 20*(5), 469–478.

18. Carlino, G., & Kerr, W. R. (2015). Agglomeration and innovation. *Handbook of Regional and Urban Economics, 5*, 349–404.

19. Wasserman, M. F., & Slack, J. D. (2020). Can there be too much specialization? Specialization in specialized courts. *Northwestern University Law Review, 151*, 1405–1503.

20. Carraro, M., Furlan, A., & Netland, T. (2024). Unlocking team performance: How shared mental models drive proactive problem-solving. *Human Relations, 78*; Stout, R. J., Cannon-Bowers, J. A., Salas, E., & Milanovich, D. M. (1999). Planning, shared mental models, and coordinated performance: An empirical link is established. *Human Factors, 41*(1), 61–71; Mathieu, J. E., Heffner, T. S., Goodwin, G. F., Salas, E., & Cannon-Bowers, J. A. (2000). The influence of shared mental models on team process and performance. *Journal of Applied Psychology, 85*(2), 273.

21. Clark, H. H. (2021). Anchoring utterances. *Topics in Cognitive Science, 13*(2), 329–350.

22. Clark, E. V. (2015). Common ground. In MacWhinney, B., & O'Grady, W. (Eds.), *The Handbook of Language Emergence* (pp. 328–353). John Wiley & Sons.

23. Bohn, M., & Köymen, B. (2018). Common ground and development. *Child Development Perspectives, 12*(2), 104–108.

24. Clark, H. H., Schreuder, R., & Buttrick, S. (1983). Common ground at the understanding of demonstrative reference. *Journal of Verbal Learning and Verbal Behavior, 22*(2), 245–258; Dombi, J., Sydorenko, T., & Timpe-Laughlin, V. (2022). Common ground, cooperation, and recipient design in human-computer interactions. *Journal of Pragmatics, 193*, 4–20.

25. Hutchins, E. (1995). *Cognition in the Wild*. MIT Press.

26. Reijula, S., & Kuorikoski, J. (2021). The diversity-ability trade-off in scientific problem solving. *Philosophy of Science, 88*(5), 894–905; Rolin, K., Koskinen, I., Kuorikoski, J., & Reijula, S. (2023). Social and cognitive diversity in science: Introduction. *Synthese, 202*(2), 36; Reijula, S., Kuorikoski, J., & MacLeod, M. (2023). The division of cognitive labor and the structure of interdisciplinary problems. *Synthese, 201*(6), 214.

27. Osiurak, F., De Oliveira, E., Navarro, J., & Reynaud, E. (2020). The castaway island: Distinct roles of theory of mind and technical reasoning in cumulative technological culture. *Journal of Experimental Psychology: General, 149*(1), 58–66.

28. Osiurak, F., Claidière, N., Bluet, A., Brogniart, J., Lasserre, S., Bonhoure, T., . . . & Reynaud, E. (2022). Technical reasoning bolsters cumulative technological culture through convergent transformations. *Science Advances, 8*(9), eabl7446.

29. Osiurak, F., Cretel, C., Duhau-Marmon, N., Fournier, I., Marignier, L., De Oliveira, E., . . . & Reynaud, E. (2020). The pedagogue, the engineer, and the friend: From whom do we learn? *Human Nature, 31*, 462–482.

30. Bluet, A., Reynaud, E., Federico, G., Bryche, C., Lesourd, M., Fournel, A., . . . & Osiurak, F. (2025). The technical-reasoning network is recruited when people observe others make or teach how to make tools: An fMRI study. *iScience, 28*(2).

31. Hartshorne, J. K., Tenenbaum, J. B., & Pinker, S. (2018). A critical period for second language acquisition: Evidence from 2/3 million English speakers. *Cognition, 177*, 263–277; Chen, T., & Hartshorne, J. K. (2021). More evidence from over 1.1 million subjects that the critical period for syntax closes in late adolescence. *Cognition, 214*, 104706.

32. Hemmatian, B., & Sloman, S. A. (2020). Two systems for thinking with a community: Outsourcing versus collaboration. In Elqayam, S., Douven, I., Evans, J., & Cruz, N. (Eds.), *Logic and Uncertainty in the Human Mind: A Tribute to David E. Over* (pp. 102–115). Routledge; Ahlstrom-Vij, K. (2016). Is there a problem with cognitive outsourcing? *Philosophical Issues, 26*; Carter, J. A. (2020). Intellectual autonomy, epistemic dependence and cognitive enhancement. *Synthese, 197*(7), 2937–2961.

33. Risko, E. F., & Gilbert, S. J. (2016). Cognitive offloading. *Trends in Cognitive Sciences, 20*(9), 676–688.

34. Gardony, A. L., Brunyé, T. T., & Taylor, H. A. (2015). Navigational aids and spatial memory impairment: The role of divided attention. *Spatial Cognition & Computation, 15*(4), 246–284; Fenech, E. P., Drews, F. A., & Bakdash, J. Z. (2010). The effects of acoustic turn-by-turn navigation on way finding. *Proceedings of the Human Factors and Ergonomics Society Annual Meeting, 54*(23); Dahmani, L., & Bohbot, V. D. (2020). Habitual use of GPS negatively impacts spatial memory during self-guided navigation. *Scientific Reports, 10*(1), 6310.

35. Henkel, L. A. (2014). Point-and-shoot memories: The influence of taking photos on memory for a museum tour. *Psychological Science, 25*(2), 396–402.

36. Wong, S. S. H., & Lim, S. W. H. (2023). Take notes, not photos: Mind-wandering mediates the impact of note-taking strategies on video-recorded lecture learning performance. *Journal of Experimental Psychology: Applied, 29*(1), 124–135.

37. Eliseev, E. D., & Marsh, E. J. (2021). Externalizing autobiographical memories in the digital age. *Trends in Cognitive Sciences, 25*(12), 1072–1081.

38. Soares, J. S., & Storm, B. C. (2018). Forget in a flash: A further investigation of the photo-taking-impairment effect. *Journal of Applied Research in Memory and Cognition, 7*(1), 154–160.

39. Eliseev, E. D., & Marsh, E. J. (2021). Externalizing autobiographical memories in the digital age. *Trends in Cognitive Sciences, 25*(12), 1072–1081.

40. Freud first described infantile amnesia in Freud, S. (1953; originally 1905). Three essays on the theory of sexuality. In Strachey, J. (Ed.), *The Standard Edition of the Complete Psychological Works of Sigmund Freud* (Vol. 7, pp. 125–248). Hogarth Press.

41. Grof, S. (1988). *The Adventure of Self-Discovery: Dimensions of Consciousness and New Perspectives in Psychotherapy and Inner Exploration.* SUNY Press; Janov, A. (1970). *The Primal Scream,* Dell Publishing. For a less charitable view see Gardner, M. (2001). Primal scream: A persistent New Age therapy. *Skeptical Inquirer, 25*(3), 17–19.

42. Yates, T. S., Fel, J., Choi, D., Trach, J. E., Behm, L., Ellis, C. T., & Turk-Browne, N. B. (2025). Hippocampal encoding of memories in human infants. *Science, 387*(6740), 1316–1320.

43. Marsh, E. J., & Rajaram, S. (2019). The digital expansion of the mind: Implications of internet usage for memory and cognition. *Journal of Applied Research in Memory and Cognition, 8*(1), 1–14.

44. Baker, N. (1994, March 27). Discards. *New Yorker,* pp. 64–86; see also Hayden, C. D. (2017). *The Card Catalog: Books, Cards, and Literary Treasures.* Chronicle Books.

45. Fisher, M., Goddu, M. K., & Keil, F. C. (2015). Searching for explanations: How the internet inflates estimates of internal knowledge. *Journal of Experimental Psychology: General, 144*(3), 674; Shlain, T. (2019). *24/6: The Power of Unplugging One Day a Week.* Simon and Schuster.

46. Huang, K. (2022, September 16). For Gen Z, TikTok is the new search engine. *New York Times.*

47. Keil, F. C., & Kominsky, J. F. (2013). Missing links in middle school: Developing use of disciplinary relatedness in evaluating internet search results. *PLOS One, 8*(6), e67777.

48. Keil, F. C. (1989). *Concepts, Kinds, and Cognitive Development.* MIT Press; Keil, F. C., & Batterman, N. (1984). A characteristic-to-defining shift in the development of word meaning. *Journal of Verbal Learning and Verbal Behavior, 23*(2), 221–236; Chi, M. T., Feltovich, P. J., & Glaser, R. (1981). Categorization and representation of physics problems by experts and novices. *Cognitive Science, 5*(2), 121–152.

49. Sparrow, B., Liu, J., & Wegner, D. M. (2011). Google effects on memory: Cognitive consequences of having information at our fingertips. *Science, 333*(6043), 776–778.

50. Fisher, Goddu, & Keil. Searching for explanations; see also Eliseev, E. D., & Marsh, E. J. (2023). Understanding why searching the internet inflates confidence in explanatory ability. *Applied Cognitive Psychology, 37*(4), 711–720.

51. Wegner, D. M., Erber, R., & Raymond, P. (1991). Transactive memory in close relationships. *Journal of Personality and Social Psychology, 61*(6), 923–929.

52. Fisher, M., & Oppenheimer, D. M. (2021). Harder than you think: How outside assistance leads to over-confidence. *Psychological Science, 32*(4), 598–610. https://doi.org/10.1177/0956797620975779

53. Fisher, M., Smiley, A. H., & Grillo, T. L. (2022). Information without knowledge: The effects of internet search on learning. *Memory, 30*(4), 375–387; Fisher, M., & Oppenheimer, D. M. (2021). Who knows what? Knowledge misattribution in the division of cognitive labor. *Journal of Experimental Psychology: Applied, 27*(2), 292.

54. Simkins, S., & Maier, M. (Eds.). (2023). *Just in Time Teaching: Across the Disciplines, and Across the Academy*. Taylor & Francis.

55. Mitchell, M. (2025). Artificial intelligence learns to reason. *Science, 387*(6740), https://www.science.org/doi/10.1126/science.adw5211

56. Cecutti, L., Chemero, A., & Lee, S. W. (2021). Technology may change cognition without necessarily harming it. *Nature Human Behaviour, 5*(8), 973–975.

57. Dicken, L., Suddendorf, T., Bulley, A., Irish, M., & Redshaw, J. (2025). Children's emerging ability to balance internal and external cognitive resources. *Child Development, 96*(2), 771–780; Gilbert, S. J., Bird, A., Carpenter, J. M., Fleming, S. M., Sachdeva, C., & Tsai, P. C. (2020). Optimal use of reminders: Metacognition, effort, and cognitive offloading. *Journal of Experimental Psychology: General, 149*(3), 501.

58. Richardson, E., & Keil, F. C. (2022). The potential for effective reasoning guides children's preference for small group discussion over crowdsourcing. *Scientific Reports, 12*(1), 1193.

59. Richardson, E., & Keil, F. C. (2022). Thinking takes time: Children use agents' response times to infer the source, quality, and complexity of their knowledge. *Cognition, 224*, 105073.

60. Thompson, C. (2014). *Smarter Than You Think: How Technology Is Changing Our Minds for the Better*. Penguin.

Chapter 9

1. Hillebrand, H. (2004). On the generality of the latitudinal diversity gradient. *American Naturalist, 163*(2), 192–211.

2. Brodie, J. F., & Mannion, P. D. (2023). The hierarchy of factors predicting the latitudinal diversity gradient. *Trends in Ecology & Evolution, 38*(1), 15–23.

3. Gaston, K. J., & Blackburn, T. M. (2000). *Pattern and Process in Macroecology*. John Wiley & Sons; Rahbek, C., Gotelli, N. J., Colwell, R. K., Entsminger, G. L., Rangel,

T. F. L., & Graves, G. R. (2007). Predicting continental-scale patterns of bird species richness with spatially explicit models. *Proceedings of the Royal Society B: Biological Sciences, 274*(1607), 165–174; Coelho, M. T. P., Barreto, E., Rangel, T. F., Diniz-Filho, J. A. F., Wüest, R. O., Bach, W., . . . & Graham, C. H. (2023). The geography of climate and the global patterns of species diversity. *Nature, 622*(7983), 537–544.

4. Horgan, J. (2015). *The End of Science: Facing the Limits of Knowledge in the Twilight of the Scientific Age.* Basic Books; Park, M., Leahey, E., & Funk, R. J. (2023). Papers and patents are becoming less disruptive over time. *Nature, 613*(7942), 138–144; Krauss, A. (2024). *Science of Science: Understanding the Foundations and Limits of Science from an Interdisciplinary Perspective.* Oxford University Press.

5. See for example Feynman, R. P. (2005). *The Pleasure of Finding Things Out: The Best Short Works of Richard P. Feynman.* Basic Books. On page 202 Feynman describes turbulence cases as commonplace and as incompatible with the elegant formalisms that are celebrated in physics.

6. Roose, K. (2023, February 16). A conversation with Bing's chatbot left me deeply unsettled. *New York Times.*

7. E.g., Ladak, A. (2024). What would qualify an artificial intelligence for moral standing? *AI and Ethics, 4*(2), 213–228; Solum, L. B. (2020). Legal personhood for artificial intelligences. In Wallach, W., & Asaro, P. (Eds.), *Machine Ethics and Robot Ethics* (pp. 415–471). Routledge.

8. Yengo, L., Vedantam, S., Marouli, E., Sidorenko, J., Bartell, E., Sakaue, S., . . . & Hirschhorn, J. N. (2022). A saturated map of common genetic variants associated with human height. *Nature, 610*, 704–712.

9. Kapoor, S., Cantrell, E. M., Peng, K., Pham, T. H., Bail, C. A., Gundersen, O. E., . . . & Narayanan, A. (2024). Reforms: Consensus-based recommendations for machine-learning-based science. *Science Advances, 10*(18), eadk3452.

10. See discussions in Shtulman, A. (2023). *Learning to Imagine: The Science of Discovering New Possibilities.* Harvard University Press; Wellman, H. M., & Gelman, S. A. (2014). Children's understanding of the nonobvious. In Sternberg, R. (Ed.), *Advances in the Psychology of Human Intelligence* (pp. 99–135). Psychology Press.

11. Taylor, J. E. T., & Taylor, G. W. (2021). Artificial cognition: How experimental psychology can help generate explainable artificial intelligence. *Psychonomic Bulletin & Review, 28*(2), 454–475; Yiu, E., Kosoy, E., & Gopnik, A. (2024). Transmission versus truth, imitation versus innovation: What children can do that large language and language-and-vision models cannot (yet). *Perspectives on Psychological Science, 19*(5), 874–883.

12. Yarkoni, T., & Westfall, J. (2017). Choosing prediction over explanation in psychology: Lessons from machine learning. *Perspectives on Psychological Science, 12*(6), 1100–1122.

13. Kapoor et al. Reforms; Hofman, J. M., Watts, D. J., Athey, S., Garip, F., Griffiths, T. L., Kleinberg, J., . . . & Yarkoni, T. (2021). Integrating explanation and prediction in computational social science. *Nature, 595*(7866), 181–188.

14. Posner, G. (2021). *Pharma: Greed, Lies, and the Poisoning of America*, p. 51. Simon and Schuster.

15. Casadevall, A. (2021). The mRNA vaccine revolution is the dividend from decades of basic science research. *Journal of Clinical Investigation, 131*(19); Callaway, E., & Naddaf, M. (2023). Pioneers of mRNA COVID vaccines win medicine Nobel. *Nature, 622*(7982), 228–229.

16. Hammack, B. (2023). *The Things We Make: The Unknown History of Invention from Cathedrals to Soda Cans*. Sourcebooks.

17. Hassan, M., Awan, F. M., Naz, A., deAndrés-Galiana, E. J., Alvarez, O., Cernea, A., . . . & Kloczkowski, A. (2022). Innovations in genomics and big data analytics for personalized medicine and health care: A review. *International Journal of Molecular Sciences, 23*(9), 4645.

18. E.g., Reichstein et al. (2019). Deep learning and process understanding for data-driven Earth system science. *Nature, 566*(7743), 195–204; Eyring, V., Collins, W. D., Gentine, P., Barnes, E. A., Barreiro, M., Beucler, T., . . . & Zanna, L. (2024). Pushing the frontiers in climate modelling and analysis with machine learning. *Nature Climate Change, 14*(9), 916–928; see also Karniadakis, G. E., Kevrekidis, I. G., Lu, L., Perdikaris, P., Wang, S., & Yang, L. (2021). Physics-informed machine learning. *Nature Reviews Physics, 3*(6), 422–440.

19. Wang, H., Fu, T., Du, Y., Gao, W., Huang, K., Liu, Z., . . . & Zitnik, M. (2023). Scientific discovery in the age of artificial intelligence. *Nature, 620*(7972), 47–60; Banerjee, S., Agarwal, A., & Singla, S. (2024). LLMs will always hallucinate, and we need to live with this. arXiv. https://doi.org/10.48550/arXiv.2409.05746; Ullman, T. (2023). Large language models fail on trivial alterations to theory-of-mind tasks. arXiv. https://doi.org/10.48550/arXiv.2409.05746; Jiang, B., Xie, Y., Hao, Z., Wang, X., Mallick, T., Su, W. J., . . . & Roth, D. (2024). A peek into token bias: Large language models are not yet genuine reasoners. arXiv. https://doi.org/10.48550/arXiv.2406.11050

20. Lawrence, E., El-Shazly, A., Seal, S., Joshi, C. K., Liò, P., Singh, S., . . . & Greenig, M. (2024). Understanding biology in the age of artificial intelligence. arXiv. https://doi.org/10.48550/arXiv.2403.04106

21. E.g., Wang, H., Fu, T., Du, Y., Gao, W., Huang, K., Liu, Z., . . . & Zitnik, M. (2023). Scientific discovery in the age of artificial intelligence. *Nature, 620*(7972), 47–60; Gao, S., Fang, A., Huang, Y., Giunchiglia, V., Noori, A., Schwarz, J. R., . . . & Zitnik, M. (2024). Empowering biomedical discovery with AI agents. *Cell, 187*(22), 6125–6151; Zheng, Z., Rampal, N., Inizan, T. J., Borgs, C., Chayes, J. T., & Yaghi, O. M.

(2025). Large language models for reticular chemistry. *Nature Reviews Materials, 10*(5), 369–381.

22. Rozenblit, L., & Keil, F. (2002). The misunderstood limits of folk science: An illusion of explanatory depth. *Cognitive Science, 26*(5), 521–562; Ylikoski, P. K. (2009). The illusion of depth of understanding in science. In de Regt, H. W., Leonelli, S., & Eigner, K. (Eds.), *Scientific Understanding: Philosophical Perspectives* (pp. 100–119), University of Pittsburgh Press; Zeveney, A. S., & Marsh, J. K. (2016). The illusion of explanatory depth in a misunderstood field: The IOED in mental disorders. In *Proceedings of the Annual Meeting of the Cognitive Science Society* (Vol. 38). https://escholarship.org/uc/item/5b65181x

23. Messeri, L., & Crockett, M. J. (2024). Artificial intelligence and illusions of understanding in scientific research. *Nature, 627*(8002), 49–58.

24. Vitriol, J. A., & Marsh, J. K. (2018). The illusion of explanatory depth and endorsement of conspiracy beliefs. *European Journal of Social Psychology, 48*(7), 955–969.

25. Goyal, A., & Bengio, Y. (2022). Inductive biases for deep learning of higher-level cognition. *Proceedings of the Royal Society A, 478*(2266), 20210068.

26. Huang, B., Low, C. J. H., Xie, F., Glymour, C., & Zhang, K. (2022). Latent hierarchical causal structure discovery with rank constraints. *Advances in Neural Information Processing Systems, 35*, 5549–5561. https://doi.org/10.48550/arXiv.2210.01798; Zhou, G., Xie, S., Hao, G., Chen, S., Huang, B., Xu, X., . . . & Zhang, K. (2023). Emerging synergies in causality and deep generative models: A survey. arXiv. https://doi.org/10.48550/arXiv.2301.12351; Sturma, N., Squires, C., Drton, M., & Uhler, C. (2024). Unpaired multi-domain causal representation learning. *Advances in Neural Information Processing Systems, 36*. https://doi.org/10.48550/arXiv.2302.00993; Li, A., Jaber, A., & Bareinboim, E. (2024). Causal discovery from observational and interventional data across multiple environments. *Advances in Neural Information Processing Systems, 36*, 16942–16956.

27. Pearl, J. (2000). *Causality: Models, Reasoning, and Inference*. Cambridge University Press.

28. Cartwright, N. (1989). *Nature's Capacities and Their Measurement*. Clarendon Press; Woodward, J. (2022). Mysteries of actual causation: It's complicated. In Willemsen, P., & Wiegmann, A. (Eds.), *Advances in Experimental Philosophy of Causation*, (chap. 2). Bloomsbury; Bareinboim, E., Correa, J. D., Ibeling, D., & Icard, T. (2022). On Pearl's hierarchy and the foundations of causal inference. In Geffner, H., Dechter, R. & Halpern, J. Y. (Eds.), *Probabilistic and Causal Inference: The Works of Judea Pearl* (pp. 507–556). ACM Books.

29. Zečević, M., Willig, M., Dhami, D. S., & Kersting, K. (2023). Causal parrots: Large language models may talk causality but are not causal. *Transactions on Machine Learning Research, 2023*(8). https://openreview.net/pdf?id=tv46tCzs83

30. Ullman, T. D., Spelke, E., Battaglia, P., & Tenenbaum, J. B. (2017). Mind games: Game engines as an architecture for intuitive physics. *Trends in Cognitive Sciences, 21*(9), 649–665; Wong, L., Grand, G., Lew, A. K., Goodman, N. D., Mansinghka, V. K., Andreas, J., & Tenenbaum, J. B. (2023). From word models to world models: Translating from natural language to the probabilistic language of thought. arXiv. https://doi.org/10.48550/arXiv.2306.12672; Sosa, F. A., Gershman, S. J., & Ullman, T. D. (2025). Blending simulation and abstraction for physical reasoning. *Cognition, 254*, 105995; Mitko, A., Navarro-Cebrián, A., Cormiea, S., & Fischer, J. (2024). A dedicated mental resource for intuitive physics. *Iscience, 27*(1).

31. Fodor, J. A. (1975). *The Language of Thought.* Harvard University Press.

32. Bohus, K. A., Cesana-Arlotti, N., Martín-Salguero, A., & Bonatti, L. L. (2023). The scope and role of deduction in infant cognition. *Current Biology, 33*(18), 4014–4020; Cesana-Arlotti, N., Kovács, Á. M., & Téglás, E. (2020). Infants recruit logic to learn about the social world. *Nature Communications, 11*(1), 5999.

33. E.g., Kaup, M., Wolff, C., Hwang, H., Mayer, J., & Bruni, E. (2024). A review of nine physics engines for reinforcement learning research. arXiv. https://doi.org/10.48550/arXiv.2407.08590; Calabrese, A., Quer, S., Squillero, G., & Tonda, A. (2024). Towards an evolutionary approach for exploiting core knowledge in artificial intelligence. In *Proceedings of the Genetic and Evolutionary Computation Conference Companion* (pp. 259–262). Association for Computing Machinery.

34. Burke, P. (2023). *Ignorance: A Global History.* Yale University Press.

35. Sequeira, A. M. (2020). Animal navigation: The mystery of open ocean orientation. *Current Biology, 30*(18), R1054–R1056.

36. DeNicola, D. R. (2018). *Understanding Ignorance: The Surprising Impact of What We Don't Know.* MIT Press.

37. Kooijman, J. D., Meijaard, J. P., Papadopoulos, J. M., Ruina, A., & Schwab, A. L. (2011). A bicycle can be self-stable without gyroscopic or caster effects. *Science, 332*(6027), 339–342.

38. When the historian Yuval Harari described "modern science" in his book *Sapiens,* he titled one chapter "The Discovery of Ignorance," and emphasized a "willingness to admit ignorance." Harari , Y. N. (2015). *Sapiens: A Brief History of Humankind.* Penguin Random House.

39. Denning, P. J., & Arquilla, J. (2022). The context problem in artificial intelligence: The artificial intelligence design challenge of teaming humans and machines is difficult because machines cannot read the context of use. *Communications of the ACM, 65*, 18–21.

40. Jiang, C., Chan, C. M., Xue, W., Liu, Q., & Guo, Y. (2025). Importance weighting can help large language models self-improve. *Proceedings of the AAAI Conference on Artificial Intelligence, 39*(23), 24257–24265.

41. Harnad, S. (2025). Language writ large: LLMs, ChatGPT, meaning, and understanding. *Frontiers in Artificial Intelligence, 7*, 1490698.

42. Oami, T., Okada, Y., & Nakada, T. A. (2024). Performance of a large language model in screening citations. *JAMA Network Open, 7*(7), e2420496–e2420496.

43. Norman, D. A. (1988). *The Psychology of Everyday Things*. Basic Books.

44. This is a Yorkshire Television interview with Richard Feynman, shown in Great Britain in 1973. Mrtp. (2015, May 28). *The world from another point of view* [video]. YouTube. https://www.youtube.com/watch?v=GNhlNSLQAFE

45. Foster, M. I., & Keane, M. T. (2015). Why some surprises are more surprising than others: Surprise as a metacognitive sense of explanatory difficulty. *Cognitive Psychology, 81*, 74–116.

46. Bayer, S., Gimpel, H., & Markgraf, M. (2022). The role of domain expertise in trusting and following explainable AI decision support systems. *Journal of Decision Systems, 32*(1), 110–138; Szymanski, M., Millecamp, M., & Verbert, K. (2021). Visual, textual or hybrid: The effect of user expertise on different explanations. In *Proceedings of the 26th International Conference on Intelligent User Interfaces* (pp. 109–119). Association for Computing Machinery.

47. Tankelevitch, L., Kewenig, V., Simkute, A., Scott, A., Sarkar, A., Sellen, A., & Rintel, S. (2023). The metacognitive demands and opportunities of generative AI. In Mueller, F., & Kyburz, P. (Eds.), *Proceedings of the 2024 CHI Conference on Human Factors in Computing Systems*. https://doi.org/10.1145/3613904.3642902

48. Johnston, A. M., Mills, C. M., & Landrum, A. R. (2015). How do children weigh competence and benevolence when deciding whom to trust? *Cognition, 144*, 76–90.

49. Rozenblit, L., & Keil, F. (2002). The misunderstood limits of folk science: An illusion of explanatory depth. *Cognitive Science, 26*(5), 521–562; Messeri & Crockett. Artificial intelligence and illusions of understanding; see also He, G., Kuiper, L., & Gadiraju, U. (2023). Knowing about knowing: An illusion of human competence can hinder appropriate reliance on AI systems. In *Proceedings of the 2023 CHI Conference on Human Factors in Computing Systems*. https://doi.org/10.1145/3544548.3581025; Byrne, R. M. (2023). Good explanations in explainable artificial intelligence (XAI): Evidence from human explanatory reasoning. In Elkind, E. (Ed.), *Proceedings of the Thirty-Second International Joint Conference on Artificial Intelligence* (pp. 6536–6544), https://doi.org/10.24963/ijcai.2023/733; Chromik, M., Eiband, M., Buchner, F., Krüger, A., & Butz, A. (2021). I think I get your point, AI! The illusion of explanatory depth in explainable AI. In *IUI '21: Proceedings of the 26th International Conference on Intelligent User Interfaces* (pp. 307–317). https://doi.org/10.1145/3397481.3450644

50. Richardson, E., Sheskin, M., & Keil, F. C. (2021). An illusion of self-sufficiency for learning about artifacts in scaffolded learners, but not observers. *Child Development*,

92(4), 1523–1538; Aboody, R., Yousif, S. R., Sheskin, M., & Keil, F. C. (2022). Says who? Children consider informants' sources when deciding whom to believe. *Journal of Experimental Psychology: General, 151*(10), 2481.

51. Yousif, S. R., Aboody, R., & Keil, F. C. (2019). The illusion of consensus: A failure to distinguish between true and false consensus. *Psychological Science, 30*(8), 1195–1204.

52. Aboody et al. Says who?, 2481.

53. Katz, R., Ogilvie, S., Shaw, J., & Woodhead, L. (2021). The art of living in a digital age. In *Gen Z, Explained*. University of Chicago Press; Hassoun, A., Beacock, I., Consolvo, S., Goldberg, B., Kelley, P. G., & Russell, D. M. (2023). Practicing information sensibility: How Gen Z engages with online information. In *Proceedings of the 2023 CHI Conference on Human Factors in Computing Systems*. https://doi.org/10.1145/3544548.3581328

54. Bertrand, A., Belloum, R., Eagan, J. R., & Maxwell, W. (2022). How cognitive biases affect XAI-assisted decision-making: A systematic review. In *Proceedings of the 2022 AAAI/ACM Conference on AI, Ethics, and Society* (pp. 78–91). https://doi.org/10.1145/3514094.3534164

55. Binz, M., & Schulz, E. (2023). Using cognitive psychology to understand GPT-3. *Proceedings of the National Academy of Sciences, 120*(6), e2218523120.

56. Liu, Z., Lee, I., Du, Y., Sanyal, S., & Zhao, J. (2023). Self-contradictory reasoning evaluation and detection. arXiv. https://doi.org/10.48550/arXiv.2311.09603

57. E.g., Gilardi, F., Alizadeh, M., & Kubli, M. (2023). ChatGPT outperforms crowd workers for text-annotation tasks. *Proceedings of the National Academy of Sciences, 120*(30), e2305016120; Webb, T., Holyoak, K. J., & Lu, H. (2023). Emergent analogical reasoning in large language models. *Nature Human Behaviour, 7*(9), 1526–1541; Aher, G. V., Arriaga, R. I., & Kalai, A. T. (2023). Using large language models to simulate multiple humans and replicate human subject studies. In Krause, A., & Brunskill, E. (Eds.), *Proceedings of the 40th International Conference on Machine Learning* (pp. 337–371); Mitchell, M., & Krakauer, D. C. (2023). The debate over understanding in AI's large language models. *Proceedings of the National Academy of Sciences, 120*(13), e2215907120; Hagendorff, T., Fabi, S., & Kosinski, M. (2023). Human-like intuitive behavior and reasoning biases emerged in large language models but disappeared in ChatGPT. *Nature Computational Science, 3*(10), 833–838; Chang, T. A., & Bergen, B. K. (2024). Language model behavior: A comprehensive survey. *Computational Linguistics, 50*(1), 293–350.

58. Jiang, G., Wang, Y., & Hui, P. (2025). Experimental exploration: Investigating cooperative interaction behavior between humans and large language model agents. arXiv. https://doi.org/10.48550/arXiv.2503.07320; Cui, Z., Li, N., & Zhou, H. (2024). Can AI replace human subjects? A large-scale replication of psychological experiments with LLMs. arXiv. https://doi.org/10.48550/arXiv.2409.00128

59. Collins, K. M., Sucholutsky, I., Bhatt, U., Chandra, K., Wong, L., Lee, M., . . . & Griffiths, T. L. (2024). Building machines that learn and think with people. *Nature Human Behaviour, 8*(10), 1851–1863; Binz, M., & Schulz, E. (2023). Turning large language models into cognitive models. arXiv. https://doi.org/10.48550/arXiv.2306.03917; Ying, L., Collins, K. M., Wong, L., Sucholutsky, I., Liu, R., Weller, A., . . . & Tenenbaum, J. B. (2025). On benchmarking human-like intelligence in machines. arXiv. https://doi.org/10.48550/arXiv.2502.20502

60. Byrne, R. M. (2023, August). Good explanations in explainable artificial intelligence (XAI): Evidence from human explanatory reasoning. In Elkind, E. (Ed.), *Proceedings of the Thirty-Second International Joint Conference on Artificial Intelligence* (pp. 6536–6544); Keane, M. T., Kenny, E. M., Delaney, E., & Smyth, B. (2021). If only we had better counterfactual explanations; Delaney, E., Pakrashi, A., Greene, D., & Keane, M. T. (2023). Counterfactual explanations for misclassified images: How human and machine explanations differ. *Artificial Intelligence, 324*, 103995; Warren, G., Keane, M. T., & Byrne, R. M. (2022). Features of explainability: How users understand counterfactual and causal explanations for categorical and continuous features in XAI. arXiv. https://doi.org/10.48550/arXiv.2204.10152

61. Delaney et al. Counterfactual explanations for misclassified images.

62. Verma, S., Boonsanong, V., Hoang, M., Hines, K., Dickerson, J., & Shah, C. (2024). Counterfactual explanations and algorithmic recourses for machine learning: A review. *ACM Computing Surveys, 56*(12), 1–42.

63. Van Noorden, R., & Perkel, J. M. (2023). AI and science: What 1,600 researchers think. *Nature, 621*(7980), 672–675.

Chapter 10

1. Talhelm, T., Zhang, X., Oishi, S., Shimin, C., Duan, D., Lan, X., & Kitayama, S. (2014). Large-scale psychological differences within China explained by rice versus wheat agriculture. *Science,* 344(6184), 603–606; Kitayama, S., & Salvador, C. E. (2024). Cultural psychology: Beyond East and West. *Annual Review of Psychology, 75*(1), 495–526.

2. Hayes, C. (2025). *The Sirens' Call: How Attention Became the World's Most Endangered Resource.* Penguin Random House.

3. E.g., Carr, N. (2025). *Superbloom: How Technologies of Connection Tear Us Apart.* W. W. Norton; Brooks, D. (2025, July 3). Are we really willing to become dumber? *New York Times.*

4. Burns, K. S. (2021). The history of social media influencers. In Watkins, B. (Ed.), *Research Perspectives on Social Media Influencers and Brand Communication* (pp. 1–22). Lexington Books.

5. Pagan, N., Mei, W., Li, C., & Dörfler, F. (2021). A meritocratic network formation model for the rise of social media influencers. *Nature Communications, 12*(1), 6865; Burns. The history of social media influencers; Tanwar, A. S., Chaudhry, H., & Srivastava, M. K. (2024). Social media influencers: Literature review, trends and research agenda. *Journal of Advances in Management Research, 21*(2), 173–202; Woolley, S. C. (2022). Digital propaganda: The power of influencers. *Journal of Democracy, 33*(3), 115–129.

6. E.g., Dweck, C. S. (2006). *Mindset: The New Psychology of Success*. Random House.

7. Epstein, D. (2021). *Range: Why Generalists Triumph in a Specialized World*. Penguin Random House.

8. Keil, F. C. (2022). *Wonder: Childhood and the Lifelong Love of Science*. MIT Press.

9. Vredenburgh, K. (2022). The right to explanation. *Journal of Political Philosophy, 30*(2), 209–229; McCarthy, D. (1997). Rights, explanation, and risks. *Ethics, 107*(2), 205–225; Taylor, E. (2024). Explanation and the right to explanation. *Journal of the American Philosophical Association, 10*(3), 467–482; Olsen, H. P., Hildebrandt, T. T., Wiesener, C., Larsen, M. S., & Flügge, A. W. A. (2024). The right to transparency in public governance: Freedom of information and the use of artificial intelligence by public agencies. *Digital Government: Research and Practice, 5*(1), 1–15; Kyrimi, E., McLachlan, S., Wohlgemut, J. M., Perkins, Z. B., Lagnado, D. A., Marsh, W., & the ExAIDSS Expert Group. (2025). Explainable AI: Definition and attributes of a good explanation for health AI. *AI Ethics, 5*, 3883–3896. https://doi.or0.1007/s43681-025-00668-x

Index

Page references followed by "t" refer to tables.

Publisher contact:
The MIT Press
Massachusetts Institute of Technology
77 Massachusetts Avenue, Cambridge, MA 02139
mitpress.mit.edu

EU Authorised Representative:
Easy Access System Europe, Mustamäe tee 50,
10621 Tallinn, Estonia
gpsr.requests@easproject.com

Printed by Integrated Books International,
United States of America